Data Mining Algorithms in C++

Data Patterns and Algorithms for Modern Applications

Timothy Masters

Apress®

Data Mining Algorithms in C++

Timothy Masters
Ithaca, New York, USA

ISBN (pbk) 978-1-4842-3314-6   ISBN (electronic) 978-1-4842-3315-3
https://doi.org/10.1007/978-1-4842-3315-3

Library of Congress Control Number: 2017962127

Cover image by Freepik (www.freepik.com)

    Managing Director: Welmoed Spahr
    Editorial Director: Todd Green
    Acquisitions Editor: Steve Anglin
    Development Editor: Matthew Moodie
    Technical Reviewers: Massimo Nardone and Michael Thomas
    Coordinating Editor: Mark Powers
    Copy Editor: Kim Wimpsett

Distributed to the book trade worldwide by Springer Science+Business Media New York, 233 Spring Street, 6th Floor, New York, NY 10013. Phone 1-800-SPRINGER, fax (201) 348-4505, e-mail orders-ny@springer-sbm.com, or visit www.springeronline.com. Apress Media, LLC is a California LLC and the sole member (owner) is Springer Science + Business Media Finance Inc (SSBM Finance Inc). SSBM Finance Inc is a **Delaware** corporation.

For information on translations, please e-mail rights@apress.com, or visit www.apress.com/rights-permissions.

Apress titles may be purchased in bulk for academic, corporate, or promotional use. eBook versions and licenses are also available for most titles. For more information, reference our Print and eBook Bulk Sales web page at www.apress.com/bulk-sales.

Any source code or other supplementary material referenced by the author in this book is available to readers on GitHub via the book's product page, located at www.apress.com/9781484233146. For more detailed information, please visit www.apress.com/source-code.

Printed on acid-free paper

Table of Contents

About the Author

Timothy Masters has a PhD in mathematical statistics with a specialization in numerical computing. He has worked predominantly as an independent consultant for government and industry. His early research involved automated feature detection in high-altitude photographs while he developed applications for flood and drought prediction, detection of hidden missile silos, and identification of threatening military vehicles. Later he worked with medical researchers in the development of computer algorithms for distinguishing between benign and malignant cells in needle biopsies. For the past 20 years he has focused primarily on methods for evaluating automated financial market trading systems. He has authored eight books on practical applications of predictive modeling.

- *Deep Belief Nets in C++ and CUDA C: Volume III: Convolutional Nets* (CreateSpace, 2016)

- *Deep Belief Nets in C++ and CUDA C: Volume II: Autoencoding in the Complex Domain* (CreateSpace, 2015)

- *Deep Belief Nets in C++ and CUDA C: Volume I: Restricted Boltzmann Machines and Supervised Feedforward Networks* (CreateSpace, 2015)

- *Assessing and Improving Prediction and Classification* (CreateSpace, 2013)

- *Neural, Novel, and Hybrid Algorithms for Time Series Prediction* (Wiley, 1995)

- *Advanced Algorithms for Neural Networks* (Wiley, 1995)

- *Signal and Image Processing with Neural Networks* (Wiley, 1994)

- *Practical Neural Network Recipes in C++* (Academic Press, 1993)

About the Technical Reviewers

Massimo Nardone has more than 23 years of experience in security, web/mobile development, cloud computing, and IT architecture. His true IT passions are security and Android.

He currently works as the chief information security officer (CISO) for Cargotec Oyj and is a member of the ISACA Finland Chapter board. Over his long career, he has held many positions including project manager, software engineer, research engineer, chief security architect, information security manager, PCI/SCADA auditor, and senior lead IT security/cloud/SCADA architect. In addition, he has been a visiting lecturer and supervisor for exercises at the Networking Laboratory of the Helsinki University of Technology (Aalto University).

Massimo has a master of science degree in computing science from the University of Salerno in Italy, and he holds four international patents (related to PKI, SIP, SAML, and proxies). Besides working on this book, Massimo has reviewed more than 40 IT books for different publishing companies and is the coauthor of *Pro Android Games* (Apress, 2015).

Michael Thomas has worked in software development for more than 20 years as an individual contributor, team lead, program manager, and vice president of engineering. Michael has more than ten years of experience working with mobile devices. His current focus is in the medical sector, using mobile devices to accelerate information transfer between patients and healthcare providers.

Introduction

Data mining is a broad, deep, and frequently ambiguous field. Authorities don't even agree on a definition for the term. What I will do is tell you how I interpret the term, especially as it applies to this book. But first, some personal history that sets the background for this book...

I've been blessed to work as a consultant in a wide variety of fields, enjoying rare diversity in my work. Early in my career, I developed computer algorithms that examined high-altitude photographs in an attempt to discover useful things. How many bushels of wheat can be expected from Midwestern farm fields this year? Are any of those fields showing signs of disease? How much water is stored in mountain ice packs? Is that anomaly a disguised missile silo? Is it a nuclear test site?

Eventually I moved on to the medical field and then finance: Does this photomicrograph of a tissue slice show signs of malignancy? Do these recent price movements presage a market collapse?

All of these endeavors have something in common: they all require that we find variables that are meaningful in the context of the application. These variables might address specific tasks, such as finding effective predictors for a prediction model. Or the variables might address more general tasks such as unguided exploration, seeking unexpected relationships among variables—relationships that might lead to novel approaches to solving the problem.

That, then, is the motivation for this book. I have taken some of my most-used techniques, those that I have found to be especially valuable in the study of relationships among variables, and documented them with basic theoretical foundations and well-commented C++ source code. Naturally, this collection is far from complete. Maybe Volume 2 will appear someday. But this volume should keep you busy for a while.

You may wonder why I have included a few techniques that are widely available in standard statistical packages, namely, very old techniques such as maximum likelihood factor analysis and varimax rotation. In these cases, I included them because they are useful, and yet reliable source code for these techniques is difficult to obtain. There are times when it's more convenient to have your own versions of old workhorses, integrated into your own personal or proprietary programs, than to be forced to coexist with canned packages that may not fetch data or present results in the way that you want.

You may want to incorporate the routines in this book into your own data mining tools. And that, in a nutshell, is the purpose of this book. I hope that you incorporate these techniques into your own data mining toolbox and find them as useful as I have in my own work.

There is no sense in my listing here the main topics covered in this text; that's what a table of contents is for. But I would like to point out a few special topics not frequently covered in other sources.

- *Information theory* is a foundation of some of the most important techniques for discovering relationships between variables, yet it is voodoo mathematics to many people. For this reason, I devote the entire first chapter to a systematic exploration of this topic. I do apologize to those who purchased my *Assessing and Improving Prediction and Classification* book as well as this one, because Chapter 1 is a nearly exact copy of a chapter in that book. Nonetheless, this material is critical to understanding much later material in this book, and I felt that it would be unfair to almost force you to purchase that earlier book in order to understand some of the most important topics in this book.

- *Uncertainty reduction* is one of the most useful ways to employ information theory to understand how knowledge of one variable lets us gain measurable insight into the behavior of another variable.

- *Schreiber's information transfer* is a fairly recent development that lets us explore causality, the directional transfer of information from one time series to another.

- *Forward stepwise selection* is a venerable technique for building up a set of predictor variables for a model. But a generalization of this method in which ranked sets of predictor candidates allow testing of large numbers of combinations of variables is orders of magnitude more effective at finding meaningful and exploitable relationships between variables.

- *Simple modifications* to relationship criteria let us detect profoundly nonlinear relationships using otherwise linear techniques.

- Now that extremely fast computers are readily available, *Monte Carlo permutation tests* are practical and broadly applicable methods for performing rigorous statistical relationship tests that until recently were intractable.

- *Combinatorially symmetric cross validation* as a means of detecting overfitting in models is a recently developed technique, which, while computationally intensive, can provide valuable information not available as little as five years ago.

- Automated selection of variables suited for predicting a given target has been routine for decades. But in many applications you have a choice of possible targets, any of which will solve your problem. Embedding target selection in the search algorithm adds a useful dimension to the development process.

- *Feature weighting as regularized energy-based learning* (FREL) is a recently developed method for ranking the predictive efficacy of a collection of candidate variables when you are in the situation of having too few cases to employ traditional algorithms.

- Everyone is familiar with *scatterplots* as a means of visualizing the relationship between pairs of variables. But they can be generalized in ways that highlight relationship anomalies far more clearly than scatterplots. Examining discrepancies between joint and marginal distributions, as well as the contribution to mutual information, in regions of the variable space can show exactly where interesting interactions are happening.

- Researchers, especially in the field of psychology, have been using *factor analysis* for decades to identify hidden dimensions in data. But few developers are aware that a frequently ignored byproduct of maximum likelihood factor analysis can be enormously useful to data miners by revealing which variables are in redundant relationships with other variables and which provide unique information.

- Everyone is familiar with using correlation statistics to measure the degree of relationship between pairs of variables, and perhaps even to extend this to the task of clustering variables that have similar behavior. But it is often the case that variables are strongly contaminated by noise, or perhaps by external factors that are not noise but that are of no interest to us. Hence, it can be useful to cluster variables *within the confines of a particular subspace* of interest, ignoring aspects of the relationships that lie outside this desired subspace.

- It is sometimes the case that a collection of time-series variables are coherent; they are impacted as a group by one or more underlying drivers, and so they change in predictable ways as time passes. Conversely, this set of variables may be mostly independent, changing on their own as time passes, regardless of what the other variables are doing. Detecting when your variables *move from one of these states* to the other allows you, among other things, to develop separate models, each optimized for the particular condition.

I have incorporated most of these techniques into a program, DATAMINE, that is available for free download, along with its user's manual. This program is not terribly elegant, as it is intended as a demonstration of the techniques presented in this book rather than as a full-blown research tool. However, the source code for its core routines that is also available for download should allow you to implement your own versions of these techniques. Please do so, and enjoy!

CHAPTER 1

Information and Entropy

Much of the material in this chapter is extracted from my prior book, Assessing and Improving Prediction and Classification. *My apologies to those readers who may feel cheated by this. However, this material is critical to the current text, and I felt that it would be unfair to force readers to buy my prior book in order to procure required background.*

The essence of data mining is the discovery of relationships among variables that we have measured. Throughout this book we will explore many ways to find, present, and capitalize on such relationships. In this chapter, we focus primarily on a specific aspect of this task: evaluating and perhaps improving the *information* content of a measured variable. What is information? This term has a rigorously defined meaning, which we will now pursue.

Entropy

Suppose you have to send a message to someone, giving this person the answer to a multiple-choice question. The catch is, you are only allowed to send the message by means of a string of ones and zeros, called *bits*. What is the minimum number of bits that you need to communicate the answer? Well, if it is a true/false question, one bit will obviously do. If four answers are possible, you will need two bits, which provide four possible patterns: 00, 01, 10, and 11. Eight answers will require three bits, and so forth. In general, to identify one of K possibilities, you will need $\log_2(K)$ bits, where $\log_2(.)$ is the logarithm base two.

Working with base-two logarithms is unconventional. Mathematicians and computer programs almost always use *natural logarithms*, in which the base is $e \approx 2.718$. The material in this chapter does not require base two; any base will do. By tradition, when natural logarithms are used in information theory, the unit of information is called

© Timothy Masters 2018
T. Masters, *Data Mining Algorithms in C++*, https://doi.org/10.1007/978-1-4842-3315-3_1

the *nat* as opposed to the *bit*. This need not concern us. For much of the remainder of this chapter, no base will be written or assumed. Any base can be used, as long as it is used consistently. Since whenever units are mentioned they will be bits, the implication is that logarithms are in base two. On the other hand, all computer programs will use natural logarithms. The difference is only one of naming conventions for the unit.

Different messages can have different worth. If you live in the midst of the Sahara Desert, a message from the weather service that today will be hot and sunny is of little value. On the other hand, a message that a foot of snow is on the way will be enormously interesting and hence valuable. A good way to quantify the value or *information* of a message is to measure the amount by which receipt of the message reduces uncertainty. If the message simply tells you something that was expected already, the message gives you little information. But if you receive a message saying that you have just won a million-dollar lottery, the message is valuable indeed and not only in the monetary sense. The fact that its information is highly unlikely gives it value.

Suppose you are a military commander. Your troops are poised to launch an invasion as soon as the order to invade arrives. All you know is that it will be one of the next 64 days, which you assume to be equally likely. You have been told that tomorrow morning you will receive a single binary message: *yes* the invasion is today or *no* the invasion is not today. Early the next morning, as you sit in your office awaiting the message, you are totally uncertain as to the day of invasion. It could be any of the upcoming 64 days, so you have six bits of uncertainty ($\log_2(64)=6$). If the message turns out to be *yes*, all uncertainty is removed. You know the day of invasion. Therefore, the information content of a *yes* message is six bits. Looked at another way, the probability of *yes* today is 1/64, so its information is $-\log_2(1/64)=6$. It should be apparent that the value of a message is inversely related to its probability.

What about a *no* message? It is certainly less valuable than *yes*, because your uncertainty about the day of invasion is only slightly reduced. You know that the invasion will not be today, which is somewhat useful, but it still could be any of the remaining 63 days. The value of *no* is $-\log_2((64-1)/64)$, which is about 0.023 bits. And yes, information in bits or nats or any other unit can be fractional.

The *expected value* of a discrete random variable on a finite set (that is, a random variable that can take on one of a finite number of different values) is equal to the sum of the product of each possible value times its probability. For example, if you have a market trading system that has a probability of winning $1,000 and a 0.6 probability of losing $500, the expected value of a trade is 0.4 * 1000 – 0.6 * 500 = $100. In the same way, we can talk about the expected value of the information content of a message. In the invasion example, the value of a *yes* message is 6 bits, and it has probability 1/64. The value of a *no* message is 0.023 bits, and its probability is 63/64. Thus, the expected value of the information in the message is (1/64) * 6 + (63/64) * 0.023 = 0.12 bits.

The invasion example had just two possible messages, *yes* and *no*. In practical applications, we will need to deal with messages that have more than two values. Consistent, rigorous notation will make it easier to describe methods for doing so. Let χ be a set that enumerates every possible message. Thus, χ may be {*yes, no*} or it may be {1, 2, 3, 4} or it may be {*benign, abnormal, malignant*} or it may be {*big loss, small loss, neutral, small win, big win*}. We will use X to generically represent a random variable that can take on values from this set, and when we observe an actual value of this random variable, we will call it x. Naturally, x will always be a member of χ. This is written as $x \varepsilon \chi$. Let $p(x)$ be the probability that x is observed. Sometimes it will be clearer to write this probability as $P(X=x)$. These two notations for the probability of observing x will be used interchangeably, depending on which is more appropriate in the context. Naturally, the sum of $p(x)$ for all $x \varepsilon \chi$ is one since χ includes every possible value of X.

Recall from the military example that the information content of a particular message x is $-\log(p(x))$, and the expected value of a random variable is the sum, across all possibilities, of its probability times its value. The information content of a message is itself a random variable. So, we can write the *expected value of the information* contained in X as shown in Equation (1.1). This quantity is called the *entropy* of X, and it is universally expressed as $H(X)$. In this equation, 0*log(0) is understood to be zero, so messages with zero probability do not contribute to entropy.

$$H(X) = -\sum_{x \varepsilon \chi} p(x) \log(p(x)) \tag{1.1}$$

Returning once more to the military example, suppose that a second message also arrives every morning: mail call. On average, mail arrives for distribution to the troops about once every three days. The actual day of arrival is random; sometimes mail will arrive several days in a row, and other times a week or more may pass with no mail. You never know when it will arrive, other than that you will be told in the morning whether mail will be delivered that day. The entropy of the *mail today* random variable is $-(1/3)$ $\log_2(1/3) - (2/3)\log_2(2/3) \approx 0.92$ bits.

In view of the fact that the entropy of the *invasion today* random variable was about 0.12 bits, this seems to be an unexpected result. How can a message that resolves an event that happens about every third day convey so much more information than one about an event that has only a 1/64 chance of happening? The answer lies in the fact that entropy is an *average*. Entropy does not measure the value of a single message. It measures the expectation of the value of the message. Even though a *yes* answer to the invasion question conveys considerable information, the fact that the nearly useless *no* message will arrive with probability 63/64 drags the average information content down to a small value.

Let K be the number of messages that are possible. In other words, the set χ contains K members. Then it can be shown (though we will not do so here) that X has maximum entropy when $p(x)=1/K$ for all $x\varepsilon\chi$. In other words, *a random variable X conveys the most information obtainable when all of its possible values are equally likely*. It is easy to see that this maximum value is $\log(K)$. Simply look at Equation (1.1) and note that all terms are equal to $(1/K)\log(1/K)$, and there are K of them. For this reason, it is often useful to observe a random variable and use Equation (1.1) to estimate its entropy and then divide this quantity by $\log(K)$ to compute its *proportional entropy*. This is a measure of how close X comes to achieving its theoretical maximum information content.

It must be noted that although the entropy of a variable is a good theoretical indicator of how much information the variable conveys, whether this information is useful is another matter entirely. Knowing whether the local post office will deliver mail today probably has little bearing on whether the home command has decided to launch an invasion today. There are ways to assess the degree to which the information content of a message is useful for making a specified decision, and these techniques will be covered later in this chapter. For now, understand that significant information content of a variable is a necessary but not sufficient condition for making effective use of that variable.

To summarize:

- Entropy is the expected value of the information contained in a variable and hence is a good measure of its potential importance.

- Entropy is given by Equation (1.1) on page 3.

- The entropy of a discrete variable is maximized when all of its possible values have equal probability.

- In many or most applications, large entropy is a necessary but not a sufficient condition for a variable to have excellent utility.

Entropy of a Continuous Random Variable

Entropy was originally defined for finite discrete random variables, and this remains its primary application. However, it can be generalized to continuous random variables. In this case, the summation of Equation (1.1) must be replaced by an integral, and the probability $p(x)$ must be replaced by the probability density function $f(x)$. The definition of entropy in the continuous case is given by Equation (1.2).

$$H(X) = -\int_{-\infty}^{\infty} f(x) \log\big(f(x)\big) dx \qquad (1.2)$$

There are several problems with continuous entropy, most of which arise from the fact that Equation (1.2) is not the limiting case of Equation (1.1) when the bin size shrinks to zero and the number of bins blows up to infinity. In practical terms, the most serious problem is that continuous entropy is not immune to rescaling. One would hope that performing the seemingly innocuous act of multiplying a random variable by a constant would leave its entropy unchanged. Intuition clearly says that it should be so because certainly the information content of a variable should be the same as the information content of ten times that variable. Alas, it is not so. Moreover, estimating a probability density function $f(x)$ from an observed sample is far more difficult than simply counting the number of observations in each of several bins for a sample. Thus, Equation (1.2) can be difficult to evaluate in applications. For these reasons, continuous entropy is avoided whenever possible. We will deal with the problem by discretizing a continuous variable in as intelligent a fashion as possible and treating the resulting random variable as discrete. The disadvantages of this approach are few, and the advantages are many.

Partitioning a Continuous Variable for Entropy

Entropy is a simple concept for discrete variables and a vile beast for continuous variables. Give me a sample of a continuous variable, and chances are I can give you a reasonable algorithm that will compute its entropy as nearly zero, an equally reasonable algorithm that will find the entropy to be huge, and any number of intermediate estimators. The bottom line is that we first need to understand our intended use for the entropy estimate and then choose an estimation algorithm accordingly.

A major use for entropy is as a screening tool for predictor variables. Entropy has theoretical value as a measure of how much information is conveyed by a variable. But it has a practical value that goes beyond this theoretical measure. There tends to be a correlation between how well many models are able to learn predictive patterns and the entropy of the predictor variables. This is not universally true, but it is true often enough that a prudent researcher will pay attention to entropy.

The mechanism by which this happens is straightforward. Many models focus their attention roughly equally across the entire range of variables, both predictor and predicted. Even models that have the theoretical capability of zooming in on important areas will have this tendency because their traditional training algorithms can require an inordinate amount of time to refocus attention onto interesting areas. The implication is that it is usually best if observed values of the variables are spread at least fairly uniformly across their range.

For example, suppose a variable has a strong right skew. Perhaps in a sample of 1,000 cases, about 900 lie in the interval 0 to 1, another 90 cases lie in 1 to 10, and the remaining 10 cases are up around 1,000. Many learning algorithms will see these few extremely large cases as providing one type of information and lump the mass of cases around zero to one into a single entity providing another type of information. The algorithm will find it difficult to identify and act on cases whose values on this variable differ by 0.1. It will be overwhelmed by the fact that some cases differ by a thousand. Some other models may do a great job of handling the mass of low-valued cases but find that the cases out in the tail are so bizarre that they essentially give up on them.

The susceptibility of models to this situation varies widely. Trees have little or no problem with skewness and heavy tails for predictors, although they have other problems that are beyond the scope of this text. Feedforward neural nets, especially those that initialize weights based on scale factors, are extremely sensitive to this condition unless trained by sophisticated algorithms. General regression neural nets and other kernel methods that use kernel widths that are relative to scale can be rendered helpless by such data. It would be a pity to come close to producing an outstanding application and be stymied by careless data preparation.

The relationship between entropy and learning is not limited to skewness and tail weight. Any unnatural clumping of data, which would usually be caught by a good entropy test, can inhibit learning by limiting the ability of the model to access information in the variable. Consider a variable whose range is zero to one. One-third of its cases lie in {0, 0.1}, one-third lie in {0.4, 0.5}, and one-third lie in {0.9, 1.0}, with output values (classes or predictions) uniformly scattered among these three clumps. This variable has no real skewness and extremely light tails. A basic test of skewness and kurtosis would show it to be ideal. Its range-to-interquartile-range ratio would be wonderful. But an entropy test would reveal that this variable is problematic. The crucial information that is crowded inside each of three tight clusters will be lost, unable to compete with the obvious difference among the three clusters. The intra-cluster variation, crucial to solving the problem, is so much less than the worthless inter-cluster variation that most models would be hobbled.

When detecting this sort of problem is our goal, the best way to partition a continuous variable is also the simplest: split the range into bins that span equal distances. Note that a technique we will explore later, splitting the range into bins containing equal numbers of cases, is worthless here. All this will do is give us an entropy of $\log(K)$, where K is the number of bins. To see why, look back at Equation (1.1) on page 3. Rather, we need to confirm that the variable in question is distributed as uniformly as possible across its range. To do this, we must split the range equally and count how many cases fall into each bin.

The code for performing this partitioning is simple; here are a few illustrative snippets. The first step is to find the range of the variable (in work here) and the factor for distributing cases into bins. Then the cases are categorized into bins. Note that two tricks are used in computing the factor. We subtract a tiny constant from the number of bins to ensure that the largest case does not overflow into a bin beyond what we have. We also add a tiny constant to the denominator to prevent division by zero in the pathological condition of all cases being identical.

```
low = high = work[0];        // Will be the variable's range
for (i=1; i<ncases; i++) {    // Check all cases to find the range
  if (work[i] > high)
    high = work[i];
  if (work[i] < low)
    low = work[i];
}
```

```
for (i=0; i<nb; i++)             // Initialize all bin counts to zero
   counts[i] = 0;

factor = (nb - 0.00000000001) / (high - low + 1.e-60);

for (i=0; i<ncases; i++) {        // Place the cases into bins
   k = (int) (factor * (work[i] - low));
   ++counts[k];
   }
```

Once the bin counts have been found, computing the entropy is a trivial application of Equation (1.1).

```
entropy = 0.0;
for (i=0; i<nb; i++) {                    // For all bins
   if (counts[i] > 0) {                   // Bin might be empty
      p = (double) counts[i] / (double) ncases;   // p(x)
      entropy -= p * log(p);              // Equation (1.1)
      }
   }

entropy /= log(nb);                       // Divide by max for proportional
```

Having a heavy tail is the most common cause of low entropy. However, clumping in the interior also appears in applications. We do need to distinguish between clumping of continuous variables due to poor design versus unavoidable grouping into discrete categories. It is the former that concerns us here. Truly discrete groups cannot be separated, while unfortunate clustering of a continuous variable can and should be dealt with. Since a heavy tail (or tails) is such a common and easily treatable occurrence and interior clumping is rarer but nearly as dangerous, it can be handy to have an algorithm that can detect undesirable interior clumping in the presence of heavy tails. Naturally, we could simply apply a transformation to lighten the tail and then perform the test shown earlier. But for quick prescreening of predictor candidates, a single test is nice to have around.

The easiest way to separate tail problems from interior problems is to dedicate one bin at each extreme to the corresponding tail. Specifically, assume that you want K bins. Find the shortest interval in the distribution that contains $(K-2)/K$ of the cases. Divide this interval into $K-2$ bins of equal width and count the number of cases in each of these interior bins. All cases below the interval go into the lowest bin. All cases above this interval go into the upper bin. If the distribution has a very long tail on one end and a very short tail on the other end, the bin on the short end may be empty. This is good because it slightly punishes the skewness. If the distribution is exactly symmetric, each of the two end bins will contain $1/K$ of the cases, which implies no penalty. This test focuses mainly on the interior of the distribution, computing the entropy primarily from the $K-2$ interior bins, with an additional small penalty for extreme skewness and no penalty for symmetric heavy tails.

Keep in mind that passing this test does not mean that we are home free. This test deliberately ignores heavy tails, so a full test must follow an interior test. Conversely, failing this interior test is bad news. Serious investigation is required.

Below, we see a code snippet that does the interior partitioning. We would follow this with the entropy calculation shown on the prior page.

```
ilow = (ncases + 1) / nb - 1;      // Unbiased lower quantile
if (ilow < 0)
   ilow = 0;

ihigh = ncases - 1 - ilow;        // Symmetric upper quantile

// Find the shortest interval containing 1-2/nbins of the distribution
qsortd (0, ncases-1, work);       // Sort cases ascending

istart = 0;                       // Beginning of interior interval
istop = istart + ihigh - ilow - 2;   // And end, inclusive
best_dist = 1.e60;                // Will be shortest distance

while (istop < ncases) {          // Try bounds containing the same n of cases
   dist = work[istop] - work[istart]; // Width of this interval

   if (dist < best_dist) {        // We're looking for the shortest
      best_dist = dist;           // Keep track of shortest
      ibest = istart;             // And its starting index
   }

   ++istart;                      // Advance to the next interval
   ++istop;                       // Keep n of cases in interval constant
}
```

```
istart = ibest;                       // This is the shortest interval
istop = istart + ihigh - ilow - 2;

counts[0] = istart;                   // The count of the leftmost bin
counts[nb-1] = ncases - istop - 1;    // and rightmost are implicit

for (i=1; i<nb-1; i++)                // Inner bins
  counts[i] = 0;

low = work[istart];                   // Lower bound of inner interval
high = work[istop];                   // And upper bound
factor = (nb - 2.00000000001) / (high - low + 1.e-60);

for (i=istart; i<=istop; i++) {       // Place cases in bins
  k = (int) (factor * (work[i] - low));
  ++counts[k+1];
  }
```

An Example of Improving Entropy

John decides that he wants to do intra-day trading of the U.S. bond futures market. One variable that he believes will be useful is an indication of how much the market is moving away from its very recent range. As a start, he subtracts from the current price a moving average of the close of the most recent 20 bars. Realizing that the importance of this deviation is relative to recent volatility, he decides to divide the price difference by the price range over those prior 20 bars. Being a prudent fellow, he does not want to divide by zero in those rare instances in which the price is flat for 20 contiguous bars, so he adds one tick (1/32 point) to the denominator. His final indicator is given by Equation (1.3).

$$X = \frac{CLOSE - MA(20)}{HIGH(20) - LOW(20) + 0.03125} \tag{1.3}$$

Being not only prudent but informed as well, he computes this indicator from a historical sample covering many years, divides the range into 20 bins, and calculates its proportional entropy as discussed on page 4. Imagine John's shock when he finds this quantity to be just 0.0027, about one-quarter of 1 percent of what should be possible! Clearly, more work is needed before this variable is presented to any prediction model.

Basic detective work reveals some fascinating numbers. The interquartile range covers −0.2 to 0.22, but the complete range is −48 to 92. There's no point in plotting a histogram; virtually the entire dataset would show up as one tall spike in the midst of a barren desert.

He now has two choices: truncate or squash. The common squashing functions, *arctangent*, *hyperbolic tangent*, and *logistic*, are all comfortable with the native domain of this variable, which happens to be about −1 to 1. Figure 1-1 shows the result of truncating this variable at +/−1. This truncated variable has a proportional entropy of 0.83, which is decent by any standard. Figure 1-2 is a histogram of the raw variable after applying the hyperbolic tangent squashing function. Its proportional entropy is 0.81. Neither approach is obviously superior, but one thing is perfectly clear: one of them, or something substantially equivalent, must be used instead of the raw variable of Equation (1.3)!

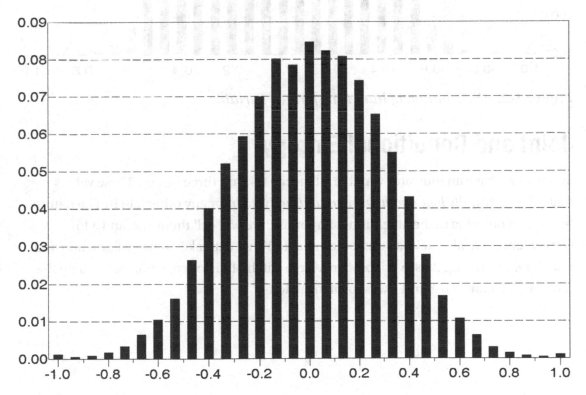

Figure 1-1. *Distribution of truncated variable*

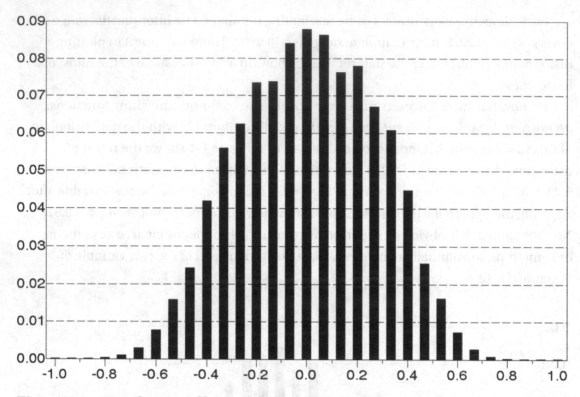

Figure 1-2. *Distribution of htan transformed variable*

Joint and Conditional Entropy

Suppose we have an indicator variable X that can take on three values. These values might be {*unusually low, about average, unusually high*} or any other labels. The nature or implied ordering of the labels is not important; we will call them 1, 2, and 3 for convenience. We also have an outcome variable Y that can take on two values: *win* and *lose*. After evaluating these variables on a large batch of historical data, we tabulate the relationship between X and Y as shown in Table 1-1.

Table 1-1. *Observed Counts and Probabilities, Theoretical Probabilities*

		Y		Marginal
		win	lose	
		80	20	100
	1	0.16	0.04	
		0.12	0.08	
		100	100	200
X	2	0.20	0.20	
		0.24	0.16	
		120	80	200
	3	0.24	0.16	
		0.24	0.16	
Marginal		300	200	500

This table shows that 80 cases fell into Category 1 of *X* and also the *win* category of *Y*, while 20 cases fell into Category 1 of *X* and also the *lose* category of *Y*, and so forth. The second number in each table cell is the fraction of all cases that fell into that cell. Thus, the (1, *win*) cell contained 0.16 of the 500 cases in the historical sample.

The third number in each cell is the fraction of cases that would, on average, fall into that cell if there were no relationship between *X* and *Y*. If two events are independent, meaning that the occurrence of one of them has no impact on the probability of occurrence of the other, the probability that they will both occur is the product of the probabilities that each will occur. In symbols, let $P(A)$ be the probability that some event *A* will occur, let $P(B)$ be the probability that some other event *B* will occur, and let $P(A,B)$ be the probability that they both will occur. Then $P(A,B)=P(A)*P(B)$ if and only if *A* and *B* are independent.

We can compute the probability of each *X* and *Y* event by summing the counts across rows and columns to get the *marginal* counts and dividing each by the total number of cases. For example, in the *Y=win* category, the total is 80+100+120=300 cases. Dividing this by 500 gives $P(Y=win)=0.6$. For *X* we find that $P(X=1)=(80+20)/500=0.2$. Hence, the probability of (*X*=1, *Y=win*), if *X* and *Y* were independent, is 0.6*0.2=0.12.

The observed probabilities for four of the six cells differ from the probabilities expected under independence, so we conclude that there might be a relationship between X and Y, though the difference is so small that random chance might just as well be responsible. An ordinary chi-square test would quantify the probability that the observed differences could have arisen from chance. But we are interested in a different approach right now.

Equation (1.1) on page 3 defined the entropy for a single random variable. We can just as well define the entropy for two random variables simultaneously. This *joint entropy* indicates how much information we obtain on average when the two variables are both known. Joint entropy is a straightforward extension of univariate entropy. Let χ, X, and x be as defined for Equation (1.1). In addition, let $¥$, Y, and y be the corresponding items for the other variable. The joint entropy $H(X, Y)$ is based on the individual cell probabilities, as shown in Equation (1.4). In this example, summing the six terms gives $H(X, Y) \approx 1.70$.

$$H(X,Y) = -\sum_{x \varepsilon \chi} \sum_{y \varepsilon ¥} p(x,y) \log(p(x,y)) \tag{1.4}$$

It often happens that the entropy of a variable is different for different values of another variable. Look back at Table 1-1. There are 100 cases for which $X=1$. Of these, 80 have $Y=win$ and 20 have $Y=lose$. The probability that $Y=win$, given that $X=1$, which is written $P(Y=win|X=1)$, is $80/100=0.8$. Similarly, $P(Y=lose|X=1)=0.2$. By Equation (1.1), the entropy of Y, given that $X=1$, which is written $H(Y|X=1)$, is $-0.8*\log(0.8) - 0.2*\log(0.2) \approx 0.50$ nats. (The switch from base 2 to base e is convenient now.) In the same way, we can compute $H(Y|X=2) \approx 0.69$, and $H(Y|X=3) \approx 0.67$.

Hold that thought. Before continuing, we need to reinforce the idea that entropy, which is a measure of disorganization, is also a measure of average information content. On the surface, this seems counterintuitive. How can it be that the more disorganized a variable is, the more information it carries? The issue is resolved if you think about what is gained by going from not knowing the value of the variable to knowing it. If the variable is highly disorganized, you gain a lot by knowing it. If you live in an area where the weather changes every hour, an accurate weather forecast (if there is such a thing) is very valuable. Conversely, if you live in the middle of a desert, a weather forecast is nearly always boring.

We just saw that we can compute the entropy of Y when X equals any specified value. This leads us to consider the entropy of Y under the general condition that we know X. In other words, we do not specify any particular X. We simply want to know, on average, what the entropy of Y will be if we happen to know X. This quantity, called the *conditional entropy of Y given X*, is an expectation once more. To compute it, we sum the product of every possibility times the probability of the possibility. In the example several paragraphs ago, we saw that $H(Y|X=1) \approx 0.50$. Looking at the marginal probabilities, we know that $P(X=1) = 100/500 = 0.20$. Following the same procedure for $X=2$ and 3, we find that the entropy of Y given that we know X, written $P(Y|X)$, is $0.2*0.50 + 0.4*0.69 + 0.4*0.67 = 0.64$.

Compare this to the entropy of Y taken alone. This is $-0.6*\log(0.6) - 0.4*\log(0.4) \approx 0.67$. Notice that the conditional entropy of Y given X is slightly less than that of Y without knowledge of X. In fact, it can be shown that $H(Y|X) \leq H(Y)$ universally. This makes sense. Knowing X certainly cannot make Y any more disorganized! If X and Y are related in any way, knowing X will reduce the disorganization of Y. Looked at another way, X may supply some of the information that would have otherwise been provided by Y. Once we know X, we have less to gain from knowing Y. A weather forecast as you roll out of bed in the morning gives you more information than the same forecast does after you have looked out the window and seen that the sky is black and rain is pouring down.

There are several standard ways of computing conditional entropy. The most straightforward way is direct application of the definition, as we did earlier. Equation (1.5) is the conditional probability of Y given X. The entropy of Y for any specified X is shown in Equation (1.6). Finally, Equation (1.7) is the entropy of Y given that we know X.

$$P(Y=y|X=x)=\frac{P(Y=y,X=x)}{P(X=x)} \tag{1.5}$$

$$H(Y|X=x)=\sum_{y\varepsilon\mathcal{Y}} P((Y=y|X=x)) \log(P((Y=y|X=x))) \tag{1.6}$$

$$H(Y|X)=\sum_{x\varepsilon\chi} P(X=x)H(Y|X=x) \tag{1.7}$$

An easier method for computing the conditional entropy of Y given X is to use the identity shown in Equation (1.8). Although the proof of this identity is simple, we will not show it here. The intuition is clear, though. The entropy of (information contained in) Y given that we already know X is the total entropy (information) minus that due strictly to X.

Rearranging the terms and treating entropy as uncertainty may make the intuition even clearer. The total uncertainty that we have about X and Y together is equal to the uncertainty we have about X plus whatever uncertainty we have about Y, given that we know X.

$$H(Y|X) = H(X,Y) - H(X) \qquad (1.8)$$

We close this section with a small exercise for you. Refer back to Table 1-1 on page 13 and look at the third line in each cell. Recall that we computed this line by multiplying the marginal probabilities. For example, $P(X=1)=100/500=0.2$, and $P(Y=win)=300/500=0.6$, which gives $0.2*0.6=0.12$ for the $(1,win)$ cell. These are the theoretical cell probabilities if X and Y were independent. Using the Y marginals, compute to decent accuracy $H(Y)$. You should get 0.673012. Using whichever formula you prefer, Equation (1.7) or (1.8), compute $H(Y|X)$ accurately. You should get the same number, 0.673012. When *theoretical* (not observed) cell probabilities are used, the entropy of Y alone is the same as the entropy of Y when X is known. Ponder why this is so.

No solid motivation for computing or examining conditional entropy is yet apparent. This will change soon. For now, let's study its computation in more detail.

Code for Conditional Entropy

The source file MUTINF_D.CPP on the Apress.com site contains a function for computing conditional entropy using the definition formula, Equation (1.7). Here are two code snippets extracted from this file. The first snippet zeros out the array where the marginal of X will be computed, and it also zeros the grid of bins that will count every combination of X and Y. It then passes through the entire dataset, filling the bins.

```
for (ix=0; ix<nbins_x; ix++) {
  marginal_x[ix] = 0;
  for (iy=0; iy<nbins_y; iy++)
    grid[ix*nbins_y+iy] = 0;
  }

for (i=0; i<ncases; i++) {
  ix = bins_x[i];
  ++marginal_x[ix];
  ++grid[ix*nbins_y+bins_y[i]];
  }
```

After the bins have been filled, the following code implements Equations (1.5) through (1.7) to compute the conditional entropy:

```
CI = 0.0;
for (ix=0; ix<nbins_x; ix++) {        // Sum Equation (1.7) for all x in X

  if (marginal_x[ix] > 0) {           // Term only makes sense if positive marginal
    cix = 0.0;                        // Will cumulate H(Y|X=x) of Equation (1.6)

    for (iy=0; iy<nbins_y; iy++) {    // Sum Equation (1.6)
      pyx = (double) grid[ix*nbins_y+iy] / (double) marginal_x[ix]; // Equation (1.5)
      if (pyx > 0.0)                  // 0 log(0) = 0
        cix += pyx * log (pyx);       // Equation (1.6)
      }
    }

  CI += cix * marginal_x[ix] / ncases; // Equation (1.7)
  }
```

Mutual Information

John has four areas of expertise: football, beer, bourbon, and poker. Mary has three areas of expertise: cooking, sewing, and poker. One night they meet at a hot game, decide that they make the perfect couple, and get married. Here are some statements about their expertise as a couple:

- John and Mary jointly have six areas of expertise: four from John, plus two from Mary (cooking, sewing) that are beyond any supplied by John. Equivalently, they have three from Mary, plus three from John (football, beer, bourbon) that are beyond any supplied by Mary. See Equation (1.9).

- John and Mary jointly have six areas of expertise: four from John, plus three from Mary, minus one (poker) that they have in common and thus was counted twice. See Equation (1.10).

- John has three areas of expertise to offer (football, beer, and bourbon) if we already have access to whatever expertise Mary offers. These three are his four, minus the one that they share. See Equation (1.11).

- Similarly, Mary has two areas of expertise above and beyond whatever is supplied by John. See Equation (1.12).

Information that is shared by two random variables X and Y is called their *mutual information*, and this quantity is written $I(X; Y)$. The following equations summarize the relationships among joint, single, and conditional entropy, and mutual information. Examination of Figure 1-3 on the next page may make the intuition behind these equations clearer.

$$H(X,Y)=H(X)+H(Y|X)=H(Y)+H(X|Y) \tag{1.9}$$

$$H(X,Y)=H(X)+H(Y)-I(X;Y) \tag{1.10}$$

$$H(X|Y)=H(X)-I(X;Y) \tag{1.11}$$

$$H(Y|X)=H(Y)-I(X;Y) \tag{1.12}$$

$$I(X;Y)=H(X)-H(X|Y)=H(Y)-H(Y|X) \tag{1.13}$$

$$I(X;Y)=H(X)+H(Y)-H(X,Y) \tag{1.14}$$

$$I(X;X)=H(X) \tag{1.15}$$

Equation (1.13) or (1.14) may be used to compute the mutual information of a pair of variables. But it is often more convenient to use the official definition of mutual information. We will not prove that the definition given by Equation (1.16) concurs with the preceding equations, as it is tedious.

$$I(X;Y)=\sum_{x \varepsilon \chi}\sum_{y \varepsilon ¥} p(x,y) \log \frac{p(x,y)}{p(x)\, p(y)} \tag{1.16}$$

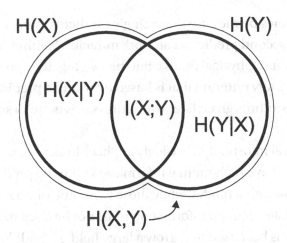

Figure 1-3. *Relationships between X and Y*

There is simple intuition behind Equation (1.16). Recall that events X and Y are independent if and only if the probability of them both happening equals the product of each of them happening: $P(X, Y)=P(X)*P(Y)$. Thus, if X and Y in Equation (1.16) are independent, the numerator will equal the denominator in the log expression. The log of one is zero, so every term in the sum will be zero. The mutual information of a pair of independent variables will evaluate to zero, as expected.

On the other hand, if X and Y have a relationship, sometimes the numerator will exceed the denominator, and sometimes it will be less. When the numerator is larger than the denominator, the log will be positive, and when the converse is true, the log will be negative. Each log term is multiplied by the numerator, with the result that positive logs will be multiplied by relatively large weights, while the negative logs will be multiplied by smaller weights. The more imbalance there is between $p(x,y)$ and $p(x)*p(y)$, the larger will be the sum.

Fano's Bound and Selection of Predictor Variables

Mutual information can be useful as a screening tool for effective predictors. It is not perfect. For one thing, mutual information picks up any sort of relationship, even unusual nonlinear dependencies. This is fine as long as the variable will be fed to a model that can take advantage of such a relationship. But naive models may be helpless, missing the information entirely. Predictive information is a necessary but not sufficient condition.

Also, it can sometimes be the case that a single predictor alone is largely useless, while pairing it with a second predictor can work miracles. Neither weight nor height alone is a good indicator of physical fitness, but the two together provide valuable information. Therefore, any criterion that is based on a single predictor variable is potentially flawed. Algorithms given later will address this issue to some degree, though not perfectly.

Nonetheless, mutual information is widely applicable as a screening tool. In general, predictor variables that have high mutual information with the predicted variable will be good candidates for use with a model, while those with little or no mutual information will make poor candidates. Mutual information must not be used to create a final set of predictors. Rather, it is best used to narrow a large field of candidates into a smaller manageable set.

In addition to the obvious intuitive value of mutual information, it has a fascinating theoretical property that can quantify its utility. [Fano, 1961, "Transmission of Information, a Statistical Theory of Communications", MIT Press.] shows that in a classification problem, the mutual information between a predictor variable and a decision variable sets a lower bound on the classification error that can be obtained. Note that there is guarantee that this accuracy can actually be realized in practice. Performance is dependent on the quality of the model being employed. Still, knowing the best that can possibly be obtained with an ideal model is useful.

Let Y be a random variable that defines a decision class from $¥=\{1, 2, ..., K\}$. In other words, there are K classes. Let X be a finite discrete random variable whose value hopefully provides information that is useful for predicting Y. Note that we are not in general asking that the value of X be the predicted value of Y. X need not even have K values. In the example of Table 1-1 on page 13, $K=2$ (*win*, *loss*), and X has three values.

We have a model that examines the value of X and predicts Y. Either this prediction is correct or it is incorrect. Let P_e be the probability that the model's prediction is in error. The *binary entropy function* is defined by Equation (1.17), and Equation (1.18) is *Fano's bound* on the attainable error of the classification model.

$$h(p) = -p \log(p) - (1-p) \log(1-p) \tag{1.17}$$

$$P_e \geq \frac{H(Y) - I(X;Y) - h(P_e)}{\log(\max(K-1,2))} \tag{1.18}$$

Officially, the denominator of Fano's bound is just $\log(K-1)$ applies only to situations in which $K>2$. To accommodate two classes, the denominator has been modified as shown earlier. Details can be found in [Erdogmus and Principe, 2003 "Insights on the Relationship Between Probability of Misclassification and Information transfer Through Classifiers." IJCSS 3:1.].

One obvious problem with Equation (1.18) is that the probability of error appears on both sides of the equation. There are two approaches to dealing with this. Sometimes we will be able to come up with a reasonable estimate of the error rate, perhaps by means of an out-of-sample test set and a good model. Then we can just blithely plug it into $h()$ in the numerator, rationalizing that the entropy and mutual information are also sample-based estimates. I've done it. In fact, I do it in one of the programs that will be presented later in this chapter. A more conservative approach is to realize that the maximum value of this term is $h(0.5)=\log(2)$. This substitution will ensure that the inequality holds, even though it will be looser than it would be if the exact value of P_e were known. Of course, if we already knew P_e, we wouldn't need the bound!

This, of course, is a valid reason for not putting much store in computed values of Fano's bound. If we already have a model in mind, any dataset that we use to compute Fano's bound gives us everything we need to compute other, probably superior, estimates of the prediction error and assorted bounds. And if we don't have a model and hence resort to using $\log(2)$ in the numerator, the bound can be overly conservative.

The real purpose of Equation (1.18) is that it alerts us to the value of the mutual information between X and Y. Mutual information is not just an obscure theoretical quantity. It plays a major role in setting a floor under the prediction accuracy that can be obtained. If we are comparing a number of candidate predictors, the denominator of Equation (1.18) will be the same for all competitors, and $H(Y)$, the entropy of the class variable, will also be constant. The error term, $h(P_e)$, may change a little, but $I(X, Y)$ is the dominant force. *The minimum attainable error rate is inversely related to the mutual information.* Therefore, candidates that have high mutual information with the class variable will probably be more useful than candidates with low mutual information.

Confusion Matrices and Mutual Information

Suppose we already have a set of predictor variables and a model that we use to predict a class. As before, Y is the true class of a case, and there are K classes. This time, we let X be the output of our model for a case. That is, X is the predicted value of Y.

Let's explore how mutual information relates to some three-by-three confusion matrices. Table 1-2 shows four examples. In each case, the row is the true class, and the column is the model's decided class. Thus, row i and column j contain the number of cases that truly belong to class i and were placed by the model in class j. Obviously, we want the diagonal to contain most cases because the diagonal represents correct classifications.

Table 1-2. *Assorted Confusion Matrices*

	4	0	6
naive	0	3	7
MI=0.173	0	0	80
	28	0	6
sure	0	26	7
MI=0.735	0	0	33
	29	2	3
spread	2	29	2
MI=0.624	2	2	29
	29	2	3
swap	2	2	29
MI=0.624	2	29	2

Mutual information quantifies a different aspect of performance than error rate. The top three confusion matrices in Table 1-2 all have an error rate of 13 percent. The first, *naive*, has very unbalanced prior probabilities. Class Three makes up 80 percent of the cases. The model takes advantage of this fact by strongly favoring this class. The result is that the other two classes are mostly misclassified. But these errors do not contribute much to the total error rate because these other two classes make up only 20 percent of cases. Mutual information easily picks up the fact that the model has not truly solved the problem. The value of 0.173 is the lowest of the set, by far.

The *sure* and *spread* confusions have identical priors (34 percent, 33 percent, 33 percent) and equal error rates, 13 percent. Yet *sure* has considerably greater mutual information than *spread*. The reason for this difference is the pattern of errors. The *spread* confusion has its errors evenly distributed among the classes, while the *sure* confusion has a consistent pattern of misclassification. Even though both models make errors at the same total rate, with the *sure* model you know in advance what sorts of errors can be expected. In particular, if the model decides that a case is in Class One or Class Two, we can be sure that the decision is correct. This knowledge of error patterns is additional information above and beyond what the error rate alone provides, and the increased mutual information reflects this fact.

Finally, look at the *swap* confusion matrix. It is identical to the *spread* confusion matrix, except that for Class Two and Class Three the model has reversed its decisions. The error rate blows up to 67 percent, while the mutual information remains at 0.624, the same as *spread*. This highlights an important property of mutual information. It is not really measuring classification performance directly. Rather, it is measuring *transfer of useful information* through the model. In other words, we are measuring one or more predictor variables and then processing these variables by a model. The variables contain some information that will be useful for making a correct decision, as well as a great deal of irrelevant information. The model acts as a filter, screening out the noise while concentrating the predictive information. The output of the model is the information that has been distilled from the predictors. The effectiveness of the model at making correct decisions is measured by its error rate. But its ability to extract useful information from a cacophony of noise is measured by its mutual information. The fact that the *swap* model has high mutual information along with a high error rate reflects the fact that the model has done a good job of finding the needles in the haystack. Its decisions really do contain useful information. The requirement that a sentient observer may be needed to process this information in a way that helps us to achieve our ultimate goal of correct classification is something that is ignored by mutual information.

Extending Fano's Bound for Upper Limits

As in the prior section, assume that we have a confusion matrix. In other words, we have a model whose output X is a prediction of the true class Y. Fano's lower bound on the error rate, shown in Equation (1.18) on page 20, can be slightly tightened if we wish. Also in this special case, we can compute an approximate upper bound on the classification error.

As was the case for the lower bound, there is little direct practical value in computing an upper bound using information theory. The data needed to compute the bound is sufficient to compute better error estimates and bounds using other methods.

However, careful study of the upper bound not only confirms the importance of mutual information as an indicator of predictive power but also yields valuable insights into effective classifier design. We will see that if we can control the way in which the classifier makes errors, we may be able to improve the theoretical limits on its true error rate.

Both the tighter lower bound and the new upper bound depend on the entropy of the error given the decision. We saw in Equation (1.18) for the lower bound that the numerator contained the binary entropy function defined in Equation (1.17). If we are willing to assume even more detailed knowledge of the pattern of errors, we can compute the conditional error entropy using Equation (1.19). In this equation, $h(.)$ is the binary entropy function of Equation (1.17), and the quantity on which it operates is the probability of error given that the model has chosen class x. Because $H(e|X)$ is less than or equal to the binary entropy of the error, the lower bound given by Equation (1.20) is tighter than that of Equation (1.18).

$$H(e|X) = \sum_{x \varepsilon \chi} P(X=x) h(P_e|X=x) \tag{1.19}$$

$$P_e \geq \frac{H(Y) - I(X;Y) - H(e|X)}{\log(\max(K-1,2))} \tag{1.20}$$

The file MUTINF_D.CPP on the Apress.com site contains a function for computing the conditional error entropy of Equation (1.19). Here is a code snippet from this file to demonstrate the computation:

```
for (ix=0; ix<nbins_x; ix++) {   // For all decision classes
   marginal_x[ix] = 0;           // Will sum marginal distribution of X
   error_count[ix] = 0;          // Will count errors associated with each decision
   }

for (i=0; i<ncases; i++) {       // Pass through all cases
   ix = bins_x[i];               // The model's decision for this case
   ++marginal_x[ix];             // Cumulate marginal distribution
   if (bins_y[i] != ix)          // If the true class is not the decision
      ++error_count[ix];         // Then this is an error, so count it
   }
```

```
CI = 0.0;                          // Will cumulate conditional error entropy here
for (ix=0; ix<nbins_x; ix++) {    // For all decision classes
   if (error_count[ix] > 0 && error_count[ix] < marginal_x[ix]) { // Avoid degenerate math
      pyx = (double) error_count[ix] / (double) marginal_x[ix];   // P(e|X=x)
      CI += (pyx * log(pyx) + (1.0-pyx) * log(1.0-pyx)) * marginal_x[ix] / ncases; // Eq 1.19
      }
   }
```

To compute an upper bound for the error rate, we need to define the conditional entropy of Y given that the model chose class x and this choice was an error. This unwieldy quantity is written as $H(Y|e, X=x)$, and it is defined by Equation (1.21). The upper bound on the error rate is then given by Equation (1.22).

$$H(Y|e,X=x)=-\sum_{y \varepsilon Y, y \neq x} \frac{P(Y=y|X=x)}{P(e|X=x)} \log\left[\frac{P(Y=y|X=x)}{P(e|X=x)}\right] \tag{1.21}$$

$$P_e \leq \frac{H(Y)-I(X;Y)-H(e|X)}{\min_x\left[H(Y|e,X=x)\right]} \tag{1.22}$$

The key fact to observe from Equation (1.22) is that the denominator is the minimum of erroneous entropy over all values of x, the predicted class. If the errors are concentrated in one or a few predicted classes, this minimum will be small, leading to a large upper bound on the theoretical error rate. This tells us that we should strive to develop a model that maximizes the entropy over all erroneous decisions, as long as we can do so without compromising the mutual information that is crucial to the numerator of the equation. In fact, the denominator of this equation is maximized (thus giving a minimum upper bound) when all errors are equiprobable.

As was stated earlier, there is little or no practical need to compute this upper bound. It is of mainly theoretical interest. But if you want to do so, code to compute the denominator of Equation (1.22), drawn from the file MUTINF_D.CPP, is as follows:

```
/*
   Compute the marginal of x and the counts in the nbins_x by nbins_y grid
*/

   for (ix=0; ix<nbins_x; ix++) {
     marginal_x[ix] = 0;
     for (iy=0; iy<nbins_y; iy++)
       grid[ix*nbins_y+iy] = 0;
     }

   for (i=0; i<ncases; i++) {
     ix = bins_x[i];
     ++marginal_x[ix];
     ++grid[ix*nbins_y+bins_y[i]];
     }

/*
   Compute the minimum entropy, conditional on error and each X Note that the computation
   in the inner loop is almost the same as in the conditional entropy. The only difference is that
   since we are also conditioning on the classification being in error, we must remove from the
   X marginal the diagonal element, which is the correct decision.
   The outer loop looks for the minimum, rather than summing.
*/

   minCI = 1.e60;
   for (ix=0; ix<nbins_x; ix++) {
     nerr = marginal_x[ix] - grid[ix*nbins_y+ix]; // Marginal that is in error
     if (nerr > 0) {
       cix = 0.0;
```

```
for (iy=0; iy<nbins_y; iy++) {
   if (iy == ix)    // This is the correct decision
      continue;    // So we exclude it; we are summing over errors
   pyx = (double) grid[ix*nbins_y+iy] / (double) nerr;    // Term in Eq 1.21
   if (pyx > 0.0)
      cix -= pyx * log (pyx);                              // Sum Eq 1.21
   }
   if (cix < minCl)
      minCl = cix;
   }
}
```

Equation (1.22) will often give an upper bound that is ridiculously excessive, sometimes much greater than one. This is especially true if $H(e|X)$ is replaced by zero in the conservative analog to how we may replace this quantity by $\log(2)$ for the lower bound. As will be vividly demonstrated in Table 1-3 on page 35, this problem is particularly severe when the denominator of Equation (1.22) is tiny because of a grossly nonuniform error distribution. In this case, we can be somewhat (though only a little) aided by the fact that a naive classifier, one that always chooses the class whose prior probability is greatest, will achieve an error rate of $1-\max_x p(x)$, where $p(x)$ is the prior probability of class x. If there are K classes and they are all equally likely, a naive classifier will have an expected error rate of $1-1/K$. If for some reason you do choose to use Equation (1.22) to compute an upper bound for the error rate, you should check it against the naive bound to be safe.

Simple Algorithms for Mutual Information

In this section we explore several of the fundamental algorithms used to compute mutual information. Later we will see how these can be modified and incorporated into sophisticated practical algorithms.

Equation (1.16) on page 18 is the standard definition of mutual information, although it is perfectly legitimate, and occasionally more efficient, to use any of the identities that preceded this equation. The file MUTINF_D.CPP contains a function that implements this definition. Here is a code snippet from this file, slightly modified for clarity:

```
/*
   Compute the marginals and the counts in the nbins_x by nbins_y grid
*/

   for (i=0; i<nbins_y; i++)
      marginal_y[i] = 0;

   for (i=0; i<nbins_x; i++) {
      marginal_x[i] = 0;
      for (j=0; j<nbins_y; j++)
         grid[i*nbins_y+j] = 0;
   }

   for (i=0; i<ncases; i++) {
      ix = bins_x[i];
      iy = bins_y[i];
      ++marginal_x[ix];
      ++marginal_y[iy];
      ++grid[ix*nbins_y+iy];
   }

/*
   Compute the mutual information
*/

   MI = 0.0;  // Will sum Eq 1.16 here

   for (i=0; i<nbins_x; i++) {
      px = (double) marginal_x[i] / (double) ncases;
```

```
for (j=0; j<nbins_y; j++) {
  py = (double) marginal_y[j] / (double) ncases;
  pxy = (double) grid[i*nbins_y+j] / (double) ncases;
  if (pxy > 0.0)
    MI += pxy * log (pxy / (px * py));   // Eq 1.16
  }
}
```

This algorithm assumes that the data is discrete. What if one or both of the variables are continuous? We saw on page 7 that the best way to partition a continuous variable *to compute its entropy* is to divide its range into bins based on equal spacing. This type of partitioning can produce unusually dense as well as unusually sparse bins, which is exactly what we want when we are estimating entropy. But *for estimating mutual information*, we would like the bin counts to reflect the relationship between the variables, rather than the marginal distributions. In the ideal situation, the marginal distribution of both variables would be uniform (all marginal bins would have equal counts) so that the counts in the grid represent the relationship between the variables to the maximum degree possible. This leads to a simple yet reasonably effective algorithm for computing the mutual information of a pair of continuous variables, or a continuous variable and a discrete variable. Later, on page 45, we will see a superior method for the case of two continuous variables. But for quick-and-dirty use or for the case of one variable being continuous and one being discrete, equal-marginal partitioning is useful.

To this end, I have an automated partitioning algorithm (source in PART.CPP) that I use in my own work. I do not guarantee that it is optimal in any particular sense, largely because there are numerous competing definitions of optimality for partitions. On the other hand, it has always behaved well for me. In particular, if you specify a desired number of bins that is at least as large as the number of different values of the variable, it will return the actual number of bins and create a single bin for each different value. Also, if the variable has few or no ties and you specify a bin count that is small relative to the number of cases, it will compute bins whose counts are approximately or exactly equal. Finally, if the variable is continuous but has numerous ties, it will group cases into bins in a way that makes sense and seems to work well. The function is called as follows:

```
void partition (
    int n,          // Input: Number of cases in the data array
    double *data,   // Input: The data array
    int *npart,     // Input/Output: Number of partitions to find; Returned as
                    // actual number of partitions, which happens if massive ties
    double *bnds,   // Output: Upper bound (inclusive) of each partition
    short int *bins // Output: Bin id (0 through npart-1) for eac h case
    )
```

The first step is to copy the data and sort it into ascending order. We need to preserve the indices of the original points, as we will need this information to assign cases to bins as the last step. Also, compute an integer array of ranks to identify ties. This is not strictly necessary, as we could simply use the floating-point data. But integer comparisons can be much faster than real comparisons on some hardware, which could make a difference for huge arrays.

```
for (i=0; i<n; i++) {
    x[i] = data[i];      // Copy the data for sorting
    indices[i] = i;      // Indices will be preserved here
    }

qsortdsi (0, n-1, x, indices);   // Sort ascending, also moving indices

ix[0] = k = 0;          // Compute ranks, including ties

for (i=1; i<n; i++) {
    if (x[i] - x[i-1] >= 1.e-12 * (1.0 + fabs(x[i]) + fabs(x[i-1])))  // Check for effective tie
        ++k;     // If not a tie, advance the counter of unique values
    ix[i] = k;
    }
```

Compute an initial set of equal-size bins, ignoring ties for now. If there are no ties, this is all we need to do.

```
k = 0;                  // Will be start of next bin up
for (i=0; i<np; i++) {  // For all partitions
  j = (n-k)/(np-i);     // Number of cases in this partition
  k += j;               // Advance the index of next one up
  bin_end[i] = k-1;     // Store upper bound of this bin
  }
```

Iteratively refine the bin boundaries until no boundary splits a tied value into different bins. Note that the upper bound of the last partition is always the last case in the sorted array, so we don't need to worry about it splitting a tie, as there are no cases above it. All we care about are the np-1 internal boundaries. Each iteration does two things. First, it removes the first splitting bound that it finds. Then it attempts to replace this lost bound by inserting a new bound in a sensible way.

```
for (;;) {              // Iterate until no ties are split across a boundary

  tie_found = 0;        // Flags if we found a split tie

  for (ibound=0; ibound<np-1; ibound++) {          // Check all boundaries
    if (ix[bin_end[ibound]] == ix[bin_end[ibound]+1]) {  // Splits a tie?

      // This bound splits a tie. Remove this bound.
      for (i=ibound+1; i<np; i++)
        bin_end[i-1] = bin_end[i];
      --np;             // We just lost a bound
      tie_found = 1;    // Flag that we found a split tie and fixed it
      break;            // Just remove one bad bound at a time
      }
    } // For all bounds, looking for a split across a tie

  if (! tie_found)      // If we got all the way through the loop
    break;              // without finding a bad bound, we are done
```

```
// The offending bound is now gone. Try splitting each remaining
// bin. For each split, check the size of the smaller resulting bin.
// Choose the split that gives the largest of the smaller.
// Note that np has been decremented, so now np < *npart.

istart = 0;
nbest = -1;

for (ibound=0; ibound<np; ibound++) {  // Check all bounds
  istop = bin_end[ibound];                 // End of this bin

  // Now processing a bin from istart through istop, inclusive
  for (i=istart; i<istop; i++) {       // Try all possible splits of this bin
    if (ix[i] == ix[i+1])                // If this splits a tie
      continue;                          // Don't check it

    nleft = i - istart + 1;             // Number of cases in left half
    nright = istop - i;                 // And right half

    if (nleft < nright) {               // If the left half is smaller
      if (nleft > nbest) {              // Keep track of the max
        nbest = nleft;                  // This is the best so far
        ibound_best = ibound;           // And its base bound
        isplit_best = i;                // Its location in the base bin
      }
    }

    else {                              // Ditto when right half is smaller
      if (nright > nbest) {
        nbest = nright;
        ibound_best = ibound;
        isplit_best = i;
      }
    }
  }

  istart = istop + 1;                   // Move on to the next bin
} // For all bounds, looking for the best bin to split
```

```
// The search is done. It may (rarely) be the case that no further
// splits are possible. This will happen if the user requests more
// partitions than there are unique values in the dataset.
// We know that this has happened if nbest is still -1. In this case
// we (obviously) cannot do a split to make up for the one lost above.

if (nbest < 0)      // If no further splits are possible
   continue;        // Then don't do it!

// We get here when the best split of an existing partition has been
// found. Save it. The bin that we are splitting is ibound_best,
// and the split for a new bound is at isplit_best.

for (ibound=np-1; ibound>=ibound_bes t; ibound--)    // Move up old bounds
   bin_end[ibound+1] = bin_end[ibound];              // To make room for new one
bin_end[ibound_best] = isplit_best;                  // The new split
++np;                                                // Count it

} // Endless search loop
```

At this point the partitioning is complete. Return the bounds to the user. Also return the bin membership of each case.

```
*npart = np; // Return the final number of partitions
for (ibound=0; ibound<np; ibound++)
   bnds[ibound] = x[bin_end[ibound]];

istart = 0;                                 // The current bin starts here
for (ibound=0; ibound<np; ibound++) {       // Process all bins
   istop = bin_end[ibound];                 // Inclusive end of this bin
   for (i=istart; i<=istop; i++)
      bins[indices[i]] = (short int) ibound;
   istart = istop + 1;
}
```

The TEST_DIS Program

The file TEST_DIS.CPP is a program that illustrates the techniques discussed so far. It allows the user to specify properties for a pair of variables, and then it generates random datasets having the specified properties and computes mutual information and some related measures. This program is for demonstration and exploration only. Later in this chapter we will present a program that reads actual datasets and processes them. The TEST_DIS program is invoked by typing its name followed by five parameters:

TEST_DIS nsamples ntries type parameter ptie

- *nsamples*: Number of cases in the dataset

- *ntries*: Number of Monte Carlo replications

- *type*: Type of test

  - 0=bivariate normal with specified correlation

  - 1=discrete bins with uniform error distribution

  - 2=discrete bins with triangular error distribution

  - 3=discrete bins with cyclic error distribution

  - 4=discrete bins with attractive class error distribution

- *parameter*: Depends on type of test

  - 0: Correlation

  - >0: Error probability

- *ptie*: If type=0, probability of a tied case, else ignored

The bivariate normal test generates two normally distributed random variables having the specified correlation. These continuous variables are partitioned into bins using the partition() subroutine presented in the prior section. All other tests generate a confusion matrix having the specified error probability. The uniform error test distributes the misclassifications to all erroneous bins with equal probability. The triangular test places most of the errors in the upper triangle. The cyclic test places the errors in a nearby class. The attractive test favors one or two unnaturally attractive classes. These all represent different types of model failure. Full details of the error distributions can be found in the source code.

A variety of numbers of bins are tested, depending on the number of cases that the user wants for each sample. The tests are repeated ntries times. For each test, it is possible to compute the theoretically correct mutual information. This enables the program to keep track of the bias and standard error of the mutual information estimates. It also computes loose and tight lower and upper bounds for misclassification error. The tight lower bounds are based on Equation (1.20) and the tight upper bounds on Equation (1.22). The loose lower bound is obtained by subtracting $h(0.5)=\log(2)$ in the numerator, as described on page 21, and the loose upper bound is obtained by not subtracting anything. The means of these bounds are computed across replications of the test. The program also counts how often the true value of the error rate falls outside the computed bounds. This demonstrates how the nature of the model's error distribution affects the width and quality of the bounds.

Table 1-3. *Some Tesults from the TEST_DIS Program*

	True	Est	Bias	StdE	| Loose |		| Tight |	
1	2.85	2.80	0.05	0.06	-0.02	0.24	0.08	0.11
2	2.88	2.84	0.04	0.04	-0.03	0.51	0.08	0.25
3	3.07	3.07	0.00	0.01	-0.09	0.66	0.02	0.11
4	3.04	3.04	0.00	0.01	-0.10	0.97	0.01	0.97

Table 1-3 shows the results from four runs of the TEST_DIS program. In all cases, 10,000 cases were in each sample. The test was replicated 1,000 times, the error rate was set at 0.1, and 32 bins were used. Observe that in all four scenarios, the estimated mutual information was very close to the true value, and the standard error of the estimate was only slightly greater than the bias, indicating that the estimates were very stable.

The loose error bounds, supposedly bounding the true value of 0.1, were universally worthless. The tight bounds were very good for the well-behaved model that had uniformly distributed errors. They deteriorated badly, though in different directions, for the triangular and cyclic error distributions. For a model with an attractive class, both the lower and the upper bounds were totally worthless. Not shown in this table is that the computed bounds never failed to enclose the true error rate.

The discussion of the TEST_DIS program is necessarily brief here. Careful study of the source code will show how the theoretical mutual information is computed, along with error bounds. Also, calling methods for the functions discussed earlier in the chapter are demonstrated.

Continuous Mutual Information

Near the beginning of this chapter we saw that entropy is fundamentally a property of finite discrete random variables, those that can take on only a finite number of fixed values. Entropy can be extended to continuous random variables by replacing summation with integration, but the continuous analog of entropy is of dubious worth in practical applications. Luckily, the situation is considerably better when it comes to mutual information. In prior sections we saw how the partition() function or something similar could be used to discretize a continuous variable into bins, and then the discrete mutual information could be computed from the bin counts. If both random variables are continuous, there are much better ways of estimating their mutual information, which is defined in Equation (1.23). (Note that if one variable is continuous and one is discrete, as would be the case when predicting a class based on a continuous predictor, the recommended procedure is to discretize the continuous variable into equal-sized bins and compute discrete mutual information.)

$$I(X;Y) = \iint f_{X,Y}(x,y) \, \log \frac{f_{X,Y}(x,y)}{f_X(x) f_Y(y)} \, dx \, dy \qquad (1.23)$$

One beautiful aspect of Equation (1.23) is that it is immune to transformations of the variables. Suppose $g(.)$ and $h(.)$ are one-to-one continuous differentiable functions over the domain of x and y, respectively. Let $x' = g(x)$ and $y' = h(y)$. Then $I(x;y) = I(x';y')$. This is in sharp contrast to continuous entropy, which is not even immune to linear rescaling, let alone nonlinear transformation.

An immensely useful corollary of this property is that observed values of the variables can be transformed to ranks or to any predefined distribution prior to computing their mutual information. This simplifies and stabilizes numerical algorithms.

The Parzen Window Method

To use Equation (1.23), we need to know the joint and marginal density functions, $f_{X,Y}(.)$, $f_X(.)$, and $f_Y(.)$. Naturally, we almost never have any knowledge of these functions other than what our data sample provides. In most cases we aren't even willing to assume a functional form such as normality. The most common way of handling this difficult situation is to use a *Parzen window approximation*.

The intuition behind a Parzen window is that areas of the domain in which the probability density is large will manifest this in the data sample by the appearance of many cases in this area. Similarly, if the probability density is small in some area of the domain, few or no cases from this area will appear in the sample. This leads to a generalized binning of the samples. Instead of defining strict boundaries for bins and counting how many cases fall into each bin, we define a weighting function, a movable window that spans the sample. When we want to compute the probability density at some point in the domain, we center the window at that point and compute a weighted sum of the cases nearby. Cases that are close to the domain point receive a large weight, while further cases receive a small weight. Very distant cases receive no weight at all. This technique is called the method of *Parzen windows*, after its inventor.

The density approximation is simple for the one-dimensional case, which covers the marginal distributions. Let the sample values be $x_1, x_2, ..., x_n$. Assume that we have a weighting function $W(d)$, which should be large when d is near zero and become smaller as d moves away from zero. Let σ be a scale factor. Then the Parzen density approximation is given by Equation (1.24).

$$f(x) = \frac{1}{n\sigma} \sum_{i=1}^{n} W\left(\frac{x - x_i}{\sigma}\right) \tag{1.24}$$

It should be clear that if the argument x has numerous cases nearby, the sum will be relatively large, because W will have many arguments near zero. Conversely, if there are no cases near x, the sum will be small, because the argument for W will be large (and hence W small) for all cases.

This is exactly what we want. The scale factor, sigma, determines the width of the window. If it is small, implying a narrow window, only cases in the immediate vicinity of x will impact the sum. If sigma is large, even distant cases will have an effect on the estimated density.

Parzen (1962) and Specht (1990a) provide a rigorous description of the properties that $W()$ must have in order for the Parzen method to be an effective density estimator. Here, we say only that these properties are reasonable: $W()$ must be bounded, go to zero rapidly as the argument goes away from zero, and integrate to unity (which is a fundamental property of a density function). The weight function favored by many is the ordinary Gaussian function of Equation (1.25).

$$W(d) = \frac{1}{\sqrt{2\pi}} e^{-d^2/2}$$

(1.25)

The Parzen density estimator is easily generalized to more than one dimension, as shown in Equations (1.26) and (1.27).

$$f(x_1, \ldots, x_p) = \frac{1}{n\sigma_1 \ldots \sigma_p} \sum_{i=1}^{n} W\left(\frac{x_1 - x_{1,i}}{\sigma_1}, \ldots, \frac{x_p - x_{p,i}}{\sigma_p}\right)$$

(1.26)

$$W(d_1, \ldots d_p) = \frac{1}{(2\pi)^{p/2}} e^{-\frac{1}{2}\sum_{1}^{p} d_i^2}$$

(1.27)

The file PARZDENS.CPP contains complete source code for computing Parzen density estimators in one, two, and three dimensions. Here we examine only a few snippets, modified for clarity when necessary, that illustrate the ideas just presented.

One aspect of the supplied code must be emphasized. Mutual information via the Parzen window method tends to be most stable when the variables have at least roughly normal distributions. For this reason, the Parzen window code applies a universal normalization transform before computing the density. (Recall that mutual information is immune to this nonlinear transformation.) The implication is that these routines *cannot* be used for general density computation. They are intended to be used only when integrating Equation (1.23), the definition of continuous mutual information. If you want to use them for other applications, you must remove the normalization code and compute the scale factor appropriately.

To estimate a normalized Parzen density in one dimension, create a ParsDens_1 object. The constructor header looks like this:

```
ParzDens_1::ParzDens_1 (
    int nd,        // Number of data points
    double *tset,  // The data array
    int div)       // Resolution divisor
```

The constructor first transforms the input data to a normal distribution. This is a standard statistical algorithm. To transform a dataset to a given distribution, first compute the cumulative distribution function (*CDF*) of the data and then map each point to the inverse CDF of the desired distribution. The sorting algorithm qsortdsi() swaps the indices along with the data.

```
for (i=0; i<nd; i++) {
    indices[i] = i;
    d[i] = tset[i];
}

qsortdsi (0, nd-1, d, indices);

for (i=0; i<nd; i++)
    d[indices[i]] = inverse_normal_cdf ((i + 1.0) / (nd + 1));
```

The sigma scale factor in Equation (1.24) is represented by std in the code. It is equal to 2.0 divided by the user's specified resolution, div. The private variable var will be used in the density computation later. The integration routine will need to know the complete practical range of the variable. Since we know that the data now follows a standard normal distribution, it is trivial to compute these limits. Finally, we compute the normalizing factor of Equations (1.24) and (1.25) so that the function integrates to unity, an essential property of a density. The code to do all this is as follows:

```
std = 2.0 / div;
var = std * std;
high = 3.0 + 3.0 * std;
low = -high;
factor = 1.0 / (nd * sqrt (2.0 * PI * var));
```

If there are numerous data points, which is the rule in practice, the summation in Equation (1.24) is slow. For this reason, the code only uses Equation (1.24) when nd is small. For large values, the constructor evaluates the density using Equation (1.24) for a reasonable number of points, and then it constructs a cubic spline interpolating function. This spline is used in future calls to the density evaluation function. Since integration involves a huge number of function calls, the savings is enormous. The spline code is tedious and uninteresting, so it will not be discussed here. See PARZDENS.CPP and SPLINE.CPP for details.

After the constructor has been called, the density (in the normalized domain, not the original domain) is estimated by calling the density() member function. Either it uses the spline approximation or it implements Equation (1.24) directly.

```
sum = 0.0;
for (i=0; i<nd; i++) {
  diff = x - d[i];
  sum += exp (-0.5 * diff * diff / var);
  }

return sum * factor;
```

The two-dimensional Parzen density code is a straightforward extension of the one-dimensional code, so it will not be shown here. It, too, uses interpolation to save time with large datasets. In this case, bilinear interpolation with quadratic extension is used. See PARZDENS.CPP and BILINEAR.CPP for details.

To compute the mutual information of a pair of variables using the Parzen window method, first create a MutualInformationParzen object. The constructor header and the most important line of code look like this:

```
MutualInformationParzen::MutualInformationParzen (
  int n,                // Number of cases
  double *depvals,   // They are here
  int div)              // Number of divisions, typically 5-10
{
dens_dep = new ParzDens_1 (n, depvals, div);
}
```

One of the two variables is supplied to the constructor. It is called depvals in the code, even though the inherent symmetry of mutual information means that there is no distinction between dependent and independent variables. The reason for this naming and for supplying one variable to the constructor is that this routine will often be used for evaluating the mutual information between a dependent variable and each of a set of candidates for independent variable. By doing as much processing as possible in the constructor, we avoid redundant computation later.

When we want to compute the mutual information between the dependent variable and a candidate predictor, the member function mutinf() is called. Its essential code, modified for clarity, is as follows:

```
this_dens_dep = dens_dep;

this_dens_trial = new ParzDens_1 (n, x, n_div);
this_dens_bivar = new ParzDens_2 (n, depvals, x, n_div);

criterion = integrate (this_dens_trial->low, this_dens_trial->high,..., outercrit);
```

The variables that start with this are statics local to the module, used to pass their data to local functions that the generic integration routine integrate() calls. This code does very little. It creates a univariate Parzen density for the candidate variable, and it creates a bivariate Parzen density for both variables. It then integrates outercrit() over the range of the candidate variable.

The real work of the algorithm is in the integration criterion routines outercrit() and innercrit(). These make up the integrand of Equation (1.23) and demonstrate a standard technique for double integration. The outer criterion, which is integrated over the range of the trial variable as shown in the prior code, itself integrates the inner criterion over the range of the dependent variable. The inner criterion needs both variables, as well as the density of the trial variable, so the two statics make it easy to pass this information from the outer criterion to the inner.

```
static double this_x, this_px; // Needed for two-dimensional integration

double outer_crit (double t)
{
  double val, high, low;
```

```
  high = this_dens_dep->high;
  low = this_dens_dep->low;
  this_x = t;
  this_px = this_dens_trial->density (this_x);
  val = integrate (low, high,..., inner_c rit);
  return val;
}

double inner_crit (double t)   // Integrand of Equation (1.23)
{
  double py, pxy, term;
  py = this_dens_dep->density (t);
  pxy = this_dens_bivar->density (t, this_x);
  term = this_px * py;    // Denominator
  if (term < 1.e-30)      // Prevent dividing by zero
    term = 1.e-30;
  term = pxy / term;      // Will take log of this
  if (term < 1.e-30)      // Prevent taking log of zero
    term = 1.e-30;
  return pxy * log (term);
}
```

The code shown here is slightly different from the code on the Apress.com site. In addition to a few changes that clarify operation, there is a difference related to the fact that the Parzen code supplied with this text converts the data to a normal distribution. Since this is the case, it is both inefficient and slightly (though not seriously) inaccurate for the inner and outer criteria to use a one-dimensional Parzen window for the marginal distributions. We already know that they are normal, so the code on the accompanying disc replaces the Parzen window with direct evaluation of the standard normal density. Comments to this effect appear in the code. This is so that the user who wants to experiment can easily switch back and forth between the two methods.

Thus far, we have conveniently pushed aside the issue of the scaling factor, sigma in Equations (1.24) and (1.26), and std in the code for the Parzen density. This is not a trivial issue. In fact, it is such a serious issue that many people avoid using Parzen windows to approximate mutual information. There are other algorithms, such as the excellent adaptive partitioning method shown in the next section. However, Parzen windows have a place in a complete toolbox. When the dataset contains just a few cases, perhaps several dozen, other methods are severely compromised. In this situation, a wide window will capture most of the important information in the distribution without running an inordinate risk of confusing random variation with true mutual information. Also, despite that an excessively wide window will bias the computed mutual information downward, while an excessively narrow window will bias it upward, this bias will be reflected nearly equally in all candidate predictors. So if the purpose of computing mutual information is to evaluate the relative quality of predictor candidates, the ranking of the candidates will be only minimally impacted by the window width, especially if the width is on the large side of optimal.

How do we choose a good window width? Ideally, we have software that plots a histogram with the Parzen density overlaid. By trying several different window widths, we can easily find the value that best captures the essence of the distribution. See, for example, Figures 1-4 through 1-7. In the absence of such a tool, a decent rule of thumb for the Parzen window software supplied with this text is to use a division factor of about five for very small samples, ten if the sample contains several hundred cases, and 15 if there are more than a thousand cases.

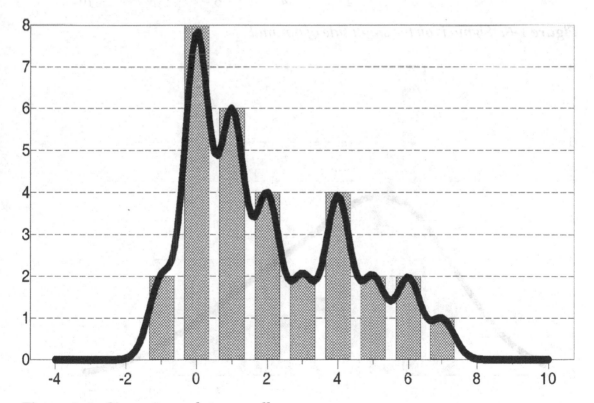

Figure 1-4. *Sigma is much too small*

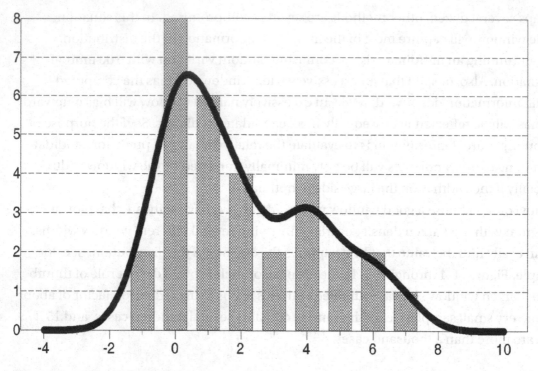

Figure 1-5. *Sigma is on the small side of optimal*

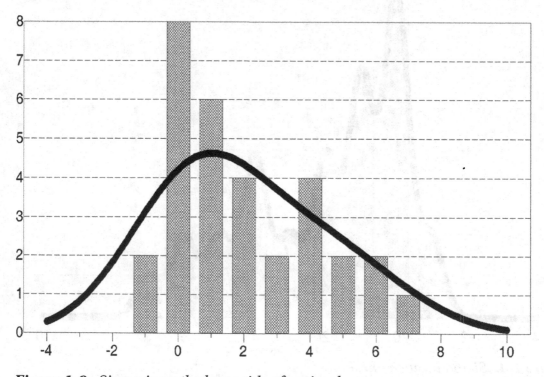

Figure 1-6. *Sigma is on the large side of optimal*

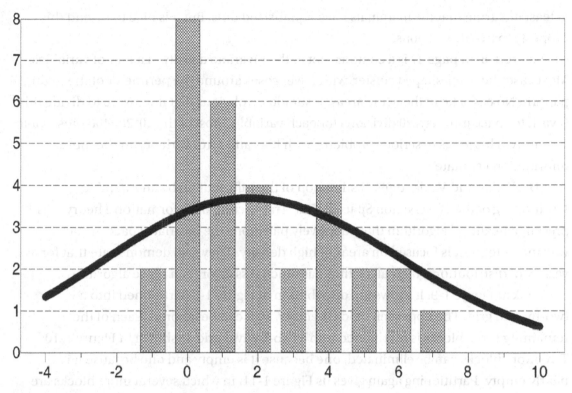

Figure 1-7. *Sigma is much too large*

Adaptive Partitioning

This section describes what is probably the best general-purpose algorithm for estimating the mutual information of two continuous variables. It is considerably more complex than the Parzen-window method just described, but the complexity is worthwhile. The algorithm is conceptually elegant and widely effective in practice. It also avoids the need to tweak a fussy parameter, which we must do for the Parzen window. It does involve two tunable parameters, but the algorithm is remarkably insensitive to their values, so in practice having to set two parameters is almost never a problem.

Recall that the naive way to compute the mutual information of a pair of continuous variables is to partition the bivariate space into a checkerboard of bins by defining boundaries for each marginal distribution and then plugging the bin counts into the discrete formula for mutual information. This was discussed on page 29. The problem with the naive method is that it pays too much attention to areas of the bivariate domain that have few or no cases, while perhaps paying too little attention to dense areas where most of the information lies. The algorithm on page 29 partially solves this problem by at least ensuring that the marginals have equal-sized bins. But it is nice to extend this property to two dimensions.

Figure 1-8 on page 47 is a contour plot of the bivariate density of a pair of variables. Most cases lie in a J-shaped cluster, with fewer cases around the perimeter of the main pattern. No cases lie in the white areas. It should be obvious that if we were to divide this bivariate space into, say, 20 divisions for each variable, most of the 20*20=400 bins would be empty. This leads to serious problems with bias and error variance in the mutual information estimate.

[Darbellay and Vajda, 1999. "Estimation of the Information by an Adaptive Partitioning of the Observation Space." IEEE Transaction on Information Theory 45:4.] present a beautiful algorithm that adaptively partitions the bivariate space in such a way that attention is focused on areas of high density. They also demonstrate that for a variety of distributions, their algorithm has much less error than naive algorithms.

Look at Figure 1-9. It shows the distribution of Figure 1-8 partitioned into a two-by-two grid. The upper-left block is empty, so it can be ignored. Each of the remaining three blocks is partitioned into a two-by-two grid as shown in Figure 1-10. Two more blocks can be eliminated, one because it is empty and one because it is nearly empty. Partitioning again gives us Figure 1-11, in which several more blocks are eliminated. It should be apparent that eventually the entire focus will be on areas of support for the density.

How far do we take the partitioning? If we stop too soon, relationships between the two variables will be obscured because details will be lost by tossing cases into overly large bins. This will downwardly bias the mutual information estimate. Conversely, if we stop too late, random variation will masquerade as actual information, inflating the estimate of the mutual information. This problem, of course, is not unique to adaptive partitioning. Anyone who experiments with the TEST_DIS program, discussed on page 34, will see it vividly displayed with naive partitioning of a bivariate normal distribution. The big difference is that since adaptive partitioning operates in two dimensions, intelligent stopping criteria are easier to implement than with naive algorithms.

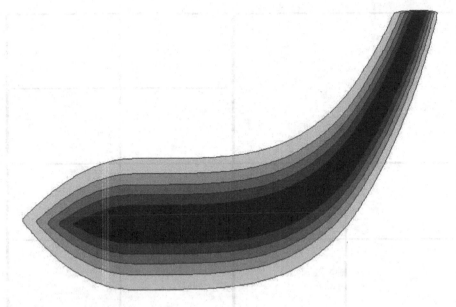

Figure 1-8. *A bivariate distribution*

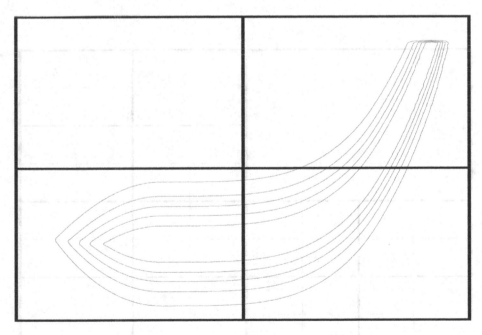

Figure 1-9. *First partitioning*

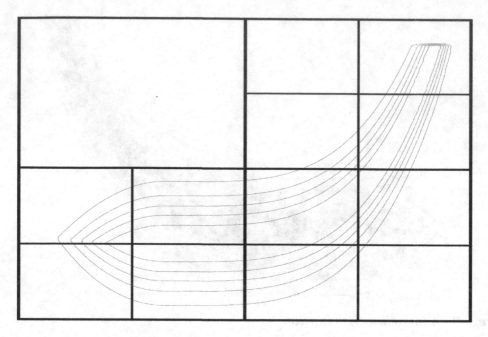

Figure 1-10. *Second partitioning*

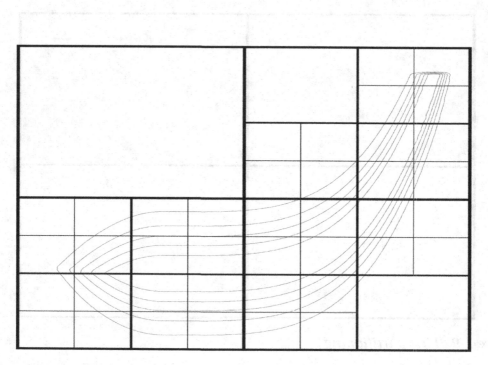

Figure 1-11. *Third partitioning*

The stopping decision is based on several tests. The first and most important is a simple chi-square test of the upcoming partition. The block whose candidacy for two-by-two subdivision is being tested is subjected to the subdivision on a trial basis. Let n_1, n_2, n_3, and n_4 be the bin counts of the four subdivisions, respectively. Let e_1, e_2, e_3, and e_4 be the expected bin counts under the null hypothesis that there is no relationship between the horizontal and vertical variables. These four expected counts will be exactly or almost exactly equal depending on whether the numbers of rows and columns are even (and hence exactly splitable in half) or odd (an exact split in half cannot be done). If the two variables are unrelated, the observed bin counts will equal the expected bin counts except for random variation. But if there is a relationship between the two variables, the counts will be skewed away from their expected values, with some bin being favored at the expense of another. The standard two-by-two chi-square test statistic is shown in Equation (1.28).

$$X^2 = \sum_{i=1}^{4} \frac{\left(\left|n_i - e_i\right| - 0.5\right)^2}{e_i} \tag{1.28}$$

If this test statistic fails to exceed the threshold for a small significance level, we conclude that the trial subdivision is probably pointless. However, it is possible that there really is a deterministic skewing of the data in the enclosing block, but a simple two-by-two subdivision fails to pick it up. This does not happen often, but it is still worth considering. For this reason, if the two-by-two chi-square test fails to detect a nonrandom distribution and if the enclosing block is relatively large, we subdivide into a four-by-four set of blocks and perform a chi-square test. If this test also fails to detect a nonrandom data distribution, we conclude that nothing is to be gained by subdividing the enclosing block, compute its contribution to the total mutual information, and henceforth ignore it.

But if either the original two-by-two chi-square test or the subsequent four-by-four test determines that the enclosing block is not uniform, we partition it into four smaller blocks. We check the size of each of these smaller blocks. If it is tiny, we compute its contribution to the total mutual information and declare that block finished. If it is still large enough for possible future splitting, we push it onto a stack of blocks to be explored and continue processing.

When a block is determined to be finished, whether because it is small or because it is uniform, its contribution to the total mutual information is computed by using a discrete approximation to Equation (1.23) on page 36. This is shown in Equation (1.29), in which p_x is the fraction of the X marginal distribution encompassed by the X dimension of the block, p_y is the fraction of the Y marginal distribution encompassed by the Y dimension of the block, and p_{xy} is the fraction of the bivariate distribution encompassed by the area of the block.

$$MI\ Contribution = p_{xy}\ \log\frac{p_{xy}}{p_x p_y} \tag{1.29}$$

We will soon present a detailed discussion of the code that implements adaptive partitioning. But since it is quite complex, we begin with a simplified statement of the algorithm. Note that the code includes an optional provision to prevent splitting across tied data. It is senseless to define a subdivision in which some cases land on one side of the trial partition while other cases whose value on the variable are equal lie on the other side. It makes more sense to place all equal values on the same side of the boundary. However, truly continuous data will never have any ties, and this provision adds to the already severe complexity of the algorithm. For these reasons, the simplified statement here will ignore ties. The topic will be covered in the discussion of the code. The algorithm is as follows:

Convert the data (*n* cases) to ranks.

Initialize *nstack*=1. This is the number of rectangles on the to-do stack. Also initialize this one stack entry to be the entire dataset. *Nstack* will be decremented when a rectangle is popped from the stack, and incremented when a rectangle is pushed onto the stack.

While *nstack* > 0 {

    Pop a rectangle from the stack

    Compute the *X* and *Y* boundaries for splitting the rectangle 2-by-2

    Compute the expected and actual bin counts in each of the four sub-rectangles

Perform a 2-by-2 chi-square test. Set the flag *splitable* to true if the test found a significant disparity in bins counts, else false.

If *splitable* = false and the rectangle is big {
    Perform a 4-by-4 chi-square test.
    If the test finds a significant disparity, set *splitable* true.
    }

If *splitable* = true {

    For each of the four sub-rectangles {

        If this rectangle is not tiny {
            Push it onto the stack
            Rearrange rectangle indices to reflect this partitioning
            }

            Else {
                Use Equation (1.29) to evaluate this sub-rectangle's contribution
                }
            }

        }

Else {
    Use Equation (1.29) to evaluate this current rectangle's contribution
    }

}

Complete code to implement the adaptive partitioning algorithm can be found in the file MUTINF_C.CPP in the accompanying code set. This code is quite complex, especially since keeping track of the nested rectangles in an efficient manner is tricky. Therefore, we will break it down into sections, slightly simplifying as needed, and discuss it one part at a time.

One of the two core components of the program is an array called indices. It is initialized to the integers 1 through n. As the algorithm progresses and rectangles are subdivided, this array will be shuffled. At any time, we can define a rectangular block by pointing to its starting and ending elements in this array. This lets us efficiently handle nesting of rectangles. For example, we may have an enclosing block that starts at element 50 of indices and ends at element 89. It may consist of four smaller blocks, defined by elements 50-59, 60-69, 70-79, and 80-89, respectively.

The other core component is a stack of rectangles to be processed. Each stack entry has the following six members:

- Xstart, Xstop: Starting and ending (inclusive) ranks of X in the rectangle

- Ystart, Ystop: Starting and ending (inclusive) ranks of Y in the rectangle

- DataStart, DataStop: Rectangle's starting and ending elements of indices

The program begins by converting each of the two variables to integer ranks. It also keeps track of tied values so that later we can avoid splitting tied cases into different partitions. Note that rather than testing for exact equality, we test for values that are nearly equal in terms of double precision. This is a good habit in most programming environments, although the reader is free to be strict if desired. Here is the code for the x variable. The other variable, y, is treated similarly.

```
for (i=0; i<n; i++) {
   work[i] = xraw[i];      // Copy the data, as we will sort it
   indices[i] = i;         // Preserve the original locations
   }

qsortdsi (0, n-1, work, indices);   // Sort ascending, also moving indices

for (i=0; i<n; i++) {
   x[indices[i]] = i;      // We now have ranks
   if (i < n-1 && work[i+1] - work[i] < 1.e-12 * (1.0 + fabs(work[i]) + fabs(work[i+1])))
     x_tied[i] = 1;        // This case is tied with one above
   else
     x_tied[i] = 0;
   }
```

To initialize, the indices array is set equal to the entire dataset, and one rectangle, the entire dataset, is placed on the to-do stack. The stack entries are inclusive, so the last index is n–1.

```
for (i=0; i<n; i++)          // For the entire dataset
   indices[i] = i;           // These are the case indices

stack[0].Xstart = 0;         // Lowest X rank in this rectangle
stack[0].Xstop = n-1;        // And highest
stack[0].Ystart = 0;         // Ditto for Y
stack[0].Ystop = n-1;
stack[0].DataStart = 0;      // Index into indices of the first case in the rectangle
stack[0].DataStop = n-1;     // And the last case
nstack = 1;                  // This is the top-of-stack pointer: One item in stack
```

The mutual information will be cumulated in MI. The program loops over the same code, processing one rectangle at a time, as long as there is at least one rectangle on the stack. The first step in the loop is to pop the rectangle off the stack.

```
MI = 0.0;                    // Will cumulate mutual information here
while (nstack > 0) {         // As long as there is a rectangle to do

   // Get the rectangle pushed onto the stack most recently
   --nstack;                                      // Pop the rectangle off the stack
   fullXstart = stack[nstack].Xstart;             // Starting X rank
   fullXstop  = stack[nstack].Xstop;              // And ending
   fullYstart = stack[nstack].Ystart;             // Ditto for Y
   fullYstop  = stack[nstack].Ystop;
   currentDataStart = stack[nstack].DataStart;    // The cases start here
   currentDataStop  = stack[nstack].DataStop;     // And end here
```

Compute the center of this rectangle in preparation for the two-by-two trial split. This center will be the rightmost (largest) index in the left (smaller rank) subrectangle. If this case happens to be tied with the next one up, we don't want to split here, as such a split would put tied cases on opposite sides of the partition. So, we set a flag to indicate whether we have this problem. If not, we are done. But if this exact center is tied, we attempt to move it off-center as little as possible, stopping as soon as we find a split that is not tied. In the pathological case that we never succeed, the tie flag remains set. We will check it later. This code is repeated for the y variable. Here we show only the x code.

```
centerX = (fullXstart + fullXstop) / 2;      // Exact center, the ideal boundary
X_AllTied = (x_tied[centerX] != 0);          // Does it happen to be tied here?
if (X_AllTied) {                             // If so, try to move it
  for (ioff=1; centerX-ioff >= fullXstart; ioff++) {   // Try to keep the offset small
    if (! x_tied[centerX-ioff]) {                // If this is not tied
      X_AllTied = 0;                         // We succeeded, so reset flag
      centerX -= ioff;                       // The new boundary is here
      break;                                 // Done searching
      }
    if (centerX + ioff == fullXstop)         // Quit if we hit the edge
      break;
    if (! x_tied[centerX+ioff]) {            // Try the other direction
      X_AllTied = 0;
      centerX += ioff;
      break;
      }
    }
  }
```

If either variable happens to be entirely tied, ideally a rare condition, the rectangle is declared to be nonsplitable. Otherwise, we trivially compute the starting and stopping indices of the four subrectangles defined by the split. The expected bin count in each partition is the total bin count times the fraction of the total x side and times the fraction of the total y side. The actual count in each partition is computed by tallying the number of cases that lie on each side of each center bound.

```
if (X_AllTied || Y_AllTied)     // If either variable is entirely tied
  splitable = 0;                // No sense trying to split
else {
```

```
trialXstart[0] = trialXstart[1] = fullXstart; // The four sub-rectangles
trialXstop[0]  = trialXstop[1]  = centerX;
trialXstart[2] = trialXstart[3] = centerX+1;
trialXstop[2]  = trialXstop[3]  = fullXstop;
trialYstart[0] = trialYstart[2] = fullYstart;
trialYstop[0]  = trialYstop[2]  = centerY;
trialYstart[1] = trialYstart[3] = centerY+1;
trialYstop[1]  = trialYstop[3]  = fullYstop;

// Compute the expected count in each of the four sub-rectangles
for (i=0; i<4; i++)
   expected[i] = (currentDataStop - currentDataStart + 1) *        // Total count
      (trialXstop[i]-trialXstart[i]+1.0) / (fullXstop-fullXstart+1.0) *   // X fraction
      (trialYstop[i]-trialYstart[i]+1.0) / (fullYstop-fullYstart+1.0);    // Y fraction

// Compute the actual count in each of the four sub-rectangles
actual[0] = actual[1] = actual[2] = actual[3] = 0;
for (i=currentDataStart; i<=currentDataStop; i++) { // All cases in this rectangle
   k = indices[i];       // Index of this case
   if (x[k] <= centerX) {      // Is it on the left side?
     if (y[k] <= centerY)     // Is it in the top half
        ++actual[0];
     else
        ++actual[1];
   }
   else {
     if (y[k] <= centerY)
        ++actual[2];
     else
        ++actual[3];
   }
}
```

Compute the two-by-two chi-square test. If the actual counts are sufficiently different from the expected counts, declare the rectangle worth splitting.

```
testval = 0.0;                    // Will cumulate test statistic here
for (i=0; i<4; i++) {             // The four sub-rectangles
   diff = fabs (actual[i] - expected[i]) - 0.5;  // Equation (1.28)
   testval += diff * diff / expected[i];
   }

splitable = (testval > chi_crit)? 1 : 0; // Does it exceed the criterion?
```

It may sometimes be the case that the rectangle really does have a nonuniform data distribution, but the cases happen to be roughly equally distributed among the four subrectangles. We can usually avoid this trap by splitting it into a four-by-four set of 16 partitions. Of course, this makes sense only if the rectangle contains more than just a few cases. I don't bother checking for ties in this finer split because it would greatly complicate the code, and this is a fairly rare occurrence anyway. The decision from the two-by-two split is the final decision the vast majority of the time. Moreover, ties will never occur in truly continuous data, so handling ties is a moot point in many or most situations.

```
if (! splitable && fullXstop-fullXstart > 30 && fullYstop-fullYstart > 30) {
    ipx = fullXstart - 1;    // Will be last index of prior sub-rectangle
    ipy = fullYstart - 1;    // Used for computing X and Y fractions
    for (i=0; i<4; i++) {    // Find the four x and y boundaries in this loop
       xcut[i] = (fullXstop - fullXstart + 1) * (i+1) / 4 + fullXstart - 1; // Rightmost limit
       xfrac[i] = (xcut[i] - ipx) / (fullXstop - fullXstart + 1.0); // Fraction in X direction
       ipx = xcut[i];                // For next pass
       ycut[i] = (fullYstop - fullYstart + 1) * (i+1) / 4 + fullYstart - 1; // Ditto for Y
       yfrac[i] = (ycut[i] - ipy) / (fullYstop - fullYstart + 1.0);
       ipy = ycut[i];
       }
```

```
// Compute expected counts
for (ix=0; ix<4; ix++) {
  for (iy=0; iy<4; iy++) {
    expected[ix*4+iy] = xfrac[ix] * yfrac[iy] *
                        (currentDataStop-currentDataStart+1);
    actual44[ix*4+iy] = 0;
    }
  }

// Compute actual counts
for (i=currentDataStart; i<=currentDataStop; i++) { // All cases in rectangle
  k = indices[i];          // Index of this case
  for (ix=0; ix<3; ix++) { // Compare x to all three inner boundaries
    if (x[k] <= xcut[ix])  // Stop before we cross incorrect boundary
      break;
    }
  for (iy=0; iy<3; iy++) { // Ditto for Y
    if (y[k] <= ycut[iy])
      break;
    }
  ++actual44[ix*4+iy];    // Tally the count
  }

// Compute the chi-square test
testval = 0.0;
for (ix=0; ix<4; ix++) {
  for (iy=0; iy<4; iy++) {
    diff = fabs (actual44[ix*4+iy] - expected[ix*4+iy]) - 0.5;
    testval += diff * diff / expected[ix*4+iy];
    }
  }
splitable = (testval > 22.0) ? 1 : 0; // Discrepancy on four-by-four test?
  } // If trying 4x4 split
} // Else not all tied
```

If the rectangle is to be split, we now process the four subrectangles. If they are not tiny, push them onto the stack for processing later. Also preserve the indices of the enclosing rectangle, because we will need them for rearranging the indices to reflect the partition.

```
if (splitable) {       // If we are to split it

  for (i=currentDataStart; i<=currentDataStop; i++)  // Preserve its indices
    current_indices[i] = indices[i];              // for rearrangement soon

  ipos = currentDataStart;        // Will rearrange indices starting here
  for (iSubRec=0; iSubRec<4; iSubRec++) { // Check all 4 sub-rectangles

    if (actual[iSubRec] >= 3) { // Big enough to push onto stack for further splitting?
      stack[nstack].Xstart = trialXstart[iSubRec];
      stack[nstack].Xstop = trialXstop[iSubRec];
      stack[nstack].Ystart = trialYstart[iSubRec];
      stack[nstack].Ystop = trialYstop[iSubRec];
      stack[nstack].DataStart = ipos;
      stack[nstack].DataStop = ipos + actual[iSubRec] - 1;
      ++nstack;
```

The current, enclosing rectangle runs from currentDataStart through currentDataStop in indices. Rearrange these indices so that the subrectangle that we just pushed has all of its cases together in a contiguous string. If we don't push any of the four, we don't need to worry about them because we will not be processing them in the future.

```
      if (iSubRec == 0) {          // Upper-left sub-rectangle
        for (i=currentDataStart; i<=currentDataStop; i++) { // All cases in rectangle
          k = current_indices[i];                    // Index of this case
          if (x[k] <= centerX && y[k] <= centerY)    // Is it in upper-left?
            indices[ipos++] = current_indices[i];    // If so, move it
        }
      }
```

```
    else if (iSubRec == 1) {
      for (i=currentDataStart; i<=currentDataStop; i++) {
        k = current_indices[i];
        if (x[k] <= centerX && y[k] > centerY)
          indices[ipos++] = current_indices[i];
        }
      }

    else if (iSubRec == 2) {
      for (i=currentDataStart; i<=currentDataStop; i++) {
        k = current_indices[i];
        if (x[k] > centerX && y[k] <= centerY)
          indices[ipos++] = current_indices[i];
        }
      }

    else { // iSubRec == 3
      for (i=currentDataStart; i<=currentDataStop; i++) {
        k = current_indices[i];
        if (x[k] > centerX && y[k] > centerY)
          indices[ipos++] = current_indices[i];
        }
      }
    } // If this sub-rectangle is large enough to be worth pushing
```

If this subrectangle is tiny, there is no reason to push it for an attempt at splitting further. Just compute its contribution to the mutual information using Equation (1.29).

```
    else { // This sub-rectangle is small, so get its contribution now
      if (actual[iSubRec] > 0) { // It only contributes if it has cases
        px = (trialXstop[iSubRec] - trialXstart[iSubRec] + 1.0) / n;
        py = (trialYstop[iSubRec] - trialYstart[iSubRec] + 1.0) / n;
        pxy = (double) actual[iSubRec] / n;
```

```
      MI += pxy * log (pxy / (px * py));  Equation (1.29)
      }
    } // Else this sub-rectangle is too small to push, so process it
   } // For all 4 sub-rectangles
 } // If splitting
```

The only other possibility is that the enclosing rectangle failed both the two-by-two and the four-by-four chi-square tests, meaning that it was so uniform that it was not worth splitting. In this case, process it using Equation (1.29).

```
else {  // Else the chi-square tests failed, so we do not split
  px = (fullXstop - fullXstart + 1.0) / n;
  py = (fullYstop - fullYstart + 1.0) / n;
  pxy = (currentDataStop - currentDataStart + 1.0) / n;
  MI += pxy * log (pxy / (px * py)); // Equation (1.29)
  }
} // While rectangles in the stack
```

This algorithm requires the user to specify only two parameters: the threshold for the two-by-two chi-square test and that for the four-by-four. The latter is so uncritical that the value 22.0 is hard-coded into the routine. The former is only slightly critical. Values between about four and eight suffice in a wide variety of circumstances. I use a value of six in all of my work, and I find this value to be universally applicable.

The TEST_CON Program

The file TEST_CON.CPP contains a complete program that demonstrates how to call the routines for using Parzen windows and adaptive partitioning to estimate mutual information for continuous variables. It also lets the user compare the performance of the two methods. The program repeatedly generates a bivariate normal dataset with specified correlation and uses both methods to estimate their mutual information. The bias and standard error of the estimates is displayed. Later in this chapter we will present a practical program for reading datasets and analyzing mutual information. The TEST_CON program is for demonstration and experimentation only. The program is invoked as follows:

TEST_CON nsamps ntries correl ptie nosplit ndiv chi

- *nsamps*: Number of cases in the dataset

- *ntries*: Number of Monte Carlo replications

- *correl*: Correlation, 0-1

- *ptie*: Probability of a tie, 0-1 (0 is generally recommended)

- *nosplit*: If nonzero, adaptive partitioning prevents splits across ties

- *ndiv*: Number of divisions for the Parzen window width

- *chi*: Two-by-two chi-square threshold for adaptive partitioning

Asymmetric Information Measures

Mutual information is symmetric in the sense that $I(X;Y) = I(Y;X)$. In other words, mutual information shows how much information two variables carry in common. This may be troubling when our goal is to use one variable, say X, to predict another, say Y. Their mutual information is based as much on the ability of Y to predict X as the ability of X to predict Y. This becomes an especially serious problem when one wants to speak of *causality*, a changing value of one variable causing a change in the probability distribution of another variable. This section will discuss two common approaches to investigating asymmetric information.

Uncertainty Reduction

Please turn back to page 19 and look at Figure 1-3, a depiction of the relationship between two variables. The two overlapping circles represent the uncertainty inherent in each variable before its value is known. Their region of overlap represents the information that is in common between them. Now suppose we have a predictor X that can take on three values, and a predicted variable Y that can take on two values. Table 1-4 shows an extreme example of asymmetric information.

Table 1-4. *Asymmetric Predictive Information*

	Y=1	Y=2
X=1	41	0
X=2	38	0
X=3	0	92

We see that there are 41 cases for which $X=1$ and $Y=1$, but no cases for which $X=1$ and $Y=2$. Examination of the other entries shows that X is a perfect predictor of Y; if we know X, then we know Y with absolute certainty. This is likely a useful thing to know about our data. But the converse is not true. When $Y=1$, our knowledge of whether X is one or two is essentially a coin toss. If our goal is to use X to predict Y, inclusion of this asymmetry in our test statistic may be counterproductive.

This can be visualized in Figure 1-3 on page 19. Call one of the entropy circles Y. Now consider how much of that circle is encompassed by the overlapping region. If the overlap encompasses most of the Y circle, then the mutual information between X and Y eliminates most of the uncertainty in Y. Conversely, if the overlap is only a small portion of the Y circle, the mutual information does little to reduce the uncertainty in Y. Note that the relationship between the overlap and the X circle (its entropy or uncertainty) plays no direct role in this computation.

This concept can be quantified by comparing the entropy of Y, which is written as $H(Y)$, with the conditional entropy of Y given that we know X, which is written as $H(Y|X)$. If these two quantities are equal, then X contributes nothing to our knowledge of Y; it has no predictive power. Conversely, if $H(Y|X)$ is zero, meaning that knowledge of X removes all uncertainty of Y, then X is a perfect predictor of Y.

The relative amount by which uncertainty in Y is reduced by knowledge of X can be expressed as shown in Equation (1.30). We have already seen the identity shown in Equation (1.31). Employing this identity in the definition gives the usual computation formula shown in Equation (1.32).

$$Uncertainty\ reduction = \frac{H(Y)-H(Y|X)}{H(Y)} \tag{1.30}$$

$$H(Y|X)=H(X,Y)-H(X) \tag{1.31}$$

$$Uncertainty\ reduction = \frac{H(X)+H(Y)-H(X,Y)}{H(Y)} \tag{1.32}$$

The file STATS.CPP provided on my web site contains a small subroutine for computing uncertainty reduction. It is listed here. Little explanation is needed because this subroutine is a direct implementation of the basic information formulas. A brief summary of its operation follows the code listing.

```cpp
void uncert_reduc (
   int nrows,           // Number of rows in data
   int ncols,           // And columns
   int *data,           // Nrows by ncols (changes fastest) matrix of cell counts
   double *row_dep,     // Returns asymmetric UR when row is dependent
   double *col_dep,     // Returns asymmetric UR when column is dependent
   double *sym,         // Returns symmetric UR
   int *rmarg,          // Work vector nrows long
   int *cmarg           // Work vector ncols long
   )
{
   int irow, icol, total;
   double p, numer, Urow, Ucol, Ujoint;

   if (nrows < 2 || ncols < 2) { // Careless user!
     *row_dep = *col_dep = *sym = 0.0;
     return;
     }

   total = 0;

   for (irow=0; irow<nrows; irow++) {
     rmarg[irow] = 0;
     for (icol=0; icol<ncols; icol++)
       rmarg[irow] += data[irow*ncols+icol];
     total += rmarg[irow];
     }
```

```
for (icol=0; icol<ncols; icol++) {
    cmarg[icol] = 0;
    for (irow=0; irow<nrows; irow++)
        cmarg[icol] += data[irow*ncols+icol];
    }

Urow = 0.0;
for (irow=0; irow<nrows; irow++) {
    if (rmarg[irow]) {
        p = (double) rmarg[irow] / (double) total;
        Urow -= p * log (p);
        }
    }

Ucol = 0.0;
for (icol=0; icol<ncols; icol++) {
    if (cmarg[icol]) {
        p = (double) cmarg[icol] / (double) total;
        Ucol -= p * log (p);
        }
    }

Ujoint = 0.0;
for (irow=0; irow<nrows; irow++) {
    for (icol=0; icol<ncols; icol++) {
        if (data[irow*ncols+icol]) {
            p = (double) data[irow*ncols+icol] / (double) total;
            Ujoint -= p * log (p);
            }
        }
    }

numer = Urow + Ucol - Ujoint;
if (Urow > 0)
    *row_dep = numer / Urow;
else
    *row_dep = 0.0;
```

```
  if (Ucol > 0)
    *col_dep = numer / Ucol;
  else
    *col_dep = 0.0;
  if (Urow + Ucol > 0)
    *sym = 2.0 * numer / (Urow + Ucol);
  else
    *sym = 0.0;
}
```

The first block of code cumulates the row marginals as well as the total case count. The second block cumulates column marginals. The next three blocks compute the row, column, and joint entropies, respectively. Finally, Equation (1.32) is used to compute the uncertainty reduction in each direction. The pooled symmetric measure computed last is not often used.

Transfer Entropy: Schreiber's Information Transfer

In 2000, Thomas Schreiber published a seminal paper on modern information theory: *Measuring Information Transfer*. His paper, [Schreiber, 2000. "Measuring Information transfer", Physical Review Letters, 85:2.], showed how we could measure a form of causality, the transfer of information from one time series to another. Later, [Vicente et al, 2011. "Transfer Entropy: A Model-Free Measure of Effective Connectivity for the Neurosciences" Journal of Computational Neuroscience 30:1.] provided some additional practical applications of Schreiber's information transfer. We now present the basic algorithm, along with code for computing information transfer (often also called *transfer entropy*).

Both of these papers discuss methods for dealing with the curse of dimensionality that plagues this computation when data is limited. These specialized algorithms come with problems of their own, and the ideal algorithm to choose is strongly application-dependent. For this reason, here we will stick with the original and most straightforward algorithm. If you are dealing with limited data and want to experiment with alternative algorithms, you should see these two papers for suggestions.

By the way, it is worth mentioning up front that the long-popular *Grainger Causality* is a special case of transfer entropy in which one assumes that the underlying model is linear autoregressive with Gaussian noise. If you are willing to accept these often restrictive assumptions, then Grainger Causality might be preferable to transfer entropy due to its more efficient use of data. However, in many applications these assumptions are too onerous to be applicable.

What is causality? Rather than digging into a deep theoretical discussion, we'll simply restate Granger's two rules:

1) The cause precedes the effect.

2) The cause contains unique information, not available in any other variable.

Note that the second rule is generally impossible to verify in practice because we cannot know for sure whether there are other variables related to the causative that we are not aware of. Still, it's nice to consider this rule in the context of an application.

To quote [Vicente et al, 2011], who in turn quotes an earlier source, "A signal X is said to cause a signal Y if the future of Y is better predicted by adding knowledge from the past and present of signal X than by using the past and present of Y alone." The code presented later shifts this back in time by one measurement period, developing the measure of causality in terms of the present value of Y being impacted by past values of X and Y. This alternate approach is more amenable to data analysis. But the traditional mathematical development that predicts future values of Y will be used in the explanations here to remain consistent with tradition. The two approaches are equivalent and differ only in starting and ending subscripts.

What we are discussing here is not the mutual information between Y and prior values of X. We might believe that this mutual information, which involves only values of X prior to the current value of Y, is a good way to quantify information transfer from X to Y. However, [Schreiber, 2000] shows that this approach has limited value and numerous problems.

An algorithm for estimating information transfer would ideally have at least the following four properties. Transfer entropy satisfies them all to a reasonable degree.

- It should not require the investigator to describe the nature of the expected interaction in advance of analysis. This property allows the algorithm to be useful for investigation.

- It should respond to common nonlinear causality modes, including purely nonlinear effects. Methods that respond only to linear components of causality, such as Granger's, are seriously limited in applicability.

- It should not be limited to just one delay for the causality. Different delays should be detectable.

- It should be reasonably robust against crosstalk, the phenomenon of a signal or noise component that appears simultaneously in X and Y. Many sources of data suffer this effect. For example, EEG measurements have common-mode noise, and equities share market-wide swings.

To rigorously present the algorithm, we need a compact notation for signifying the current and recent historical values of a time series. In particular, at time t we will represent the k most recent values of X (including the current value) as $X_t^{(K)} = (X_t, X_{t-1}, \ldots, X_{t-k+1})$, and similarly for Y.

We also need a brief detour to discuss the *Kullback-Liebler distance* between two discrete probability distributions. Suppose P and Q are discrete probability distributions over some domain indicated by i. Then the Kullback-Liebler distance between P and Q is given by Equation (1.33).

$$D(P \| Q) = \sum_i p(i) \log\left(\frac{p(i)}{q(i)} \right)$$
(1.33)

A little intuition about this definition is in order. Suppose, for example, that the two distributions are identical. In other words, the probability of every possible event is the same in both distributions. In this case, the ratio will be one for every i, and the log of one is zero. So the K-L distance will be zero. Now suppose that for some event the probability under P of that event is much larger than under Q. The ratio is greater than one, so the log will be positive, and the weight will be unusually large, resulting in a large contribution to the sum. Conversely, suppose for some event its probability under Q is much larger than its probability under P. Now the ratio will be less than one, the log will be negative, but the weight will be small, so only a small value will be subtracted from the sum. The more the two distributions diverge, the greater will be the sum.

We state without proof that this sum can never be negative, which is a nice property for a distance! But it is not symmetric: $D(P \| Q)$ does not necessarily equal $D(Q \| P)$. Rather, the *K-L* distance measures the amount of information lost when the distribution Q is used to approximate P. In most applications, P is the (assumed) true distribution of the data, while Q is some experimental approximation of P, perhaps based on a proposed model or other tentative explanation of P.

We are now ready to proceed. Recall that we know current and historical values of Y, and this knowledge gives us some ability to predict the next value of Y. Our goal in computing information transfer is to measure the degree to which the additional knowledge of current and historical values of X adds to our ability to predict the next Y. Equivalently, we will measure the amount of predictive information that is lost by denying ourselves knowledge of X.

Suppose we are at observation time t. If we have knowledge of the historical values of both X and Y, then we can write the probability of the next $(t+1)$ value of Y as $p(y_{t+1}|y_t^{(n)}, x_t^{(m)})$, where n and m may be different (we may know different lengths of X and Y history). But if we do not know X, then the probability of the next value of Y is $p(y_{t+1}|y_t^{(n)})$. If X has no causative effect on Y, then these two probabilities are equal for all possible outcomes. But if X does have causative effect, then they will differ.

We are now in a position to define transfer entropy. Recall that the Kullback-Liebler distance $D(P \parallel Q)$ measures the amount of information lost when the distribution Q is used to approximate P. The actually observed data provides $p(y_{t+1}|y_t^{(n)}, x_t^{(m)})$. What if we were to approximate this with the probability distribution that lacks access to X, namely, $p(y_{t+1}|y_t^{(n)})$? The former plays the role of P, and the latter plays the role of Q. Because of the conditional probabilities, we must sum across the conditions. The information lost by denying knowledge of X is the transfer entropy from X to Y, and it is defined as shown in Equation (1.34).

$$Transfer\ entropy = \sum p\left(y_{t+1}, y_t^{(n)}, x_t^{(m)}\right) \log\left(\frac{p\left(y_{t+1}|y_t^{(n)}, x_t^{(m)}\right)}{p\left(y_{t+1}|y_t^{(n)}\right)}\right) \qquad (1.34)$$

We can define the required conditional probabilities in terms of primitive probabilities, shown here using our current notation:

$$p\left(y_{t+1}|y_t^{(n)}, x_t^{(m)}\right) = \frac{p\left(y_{t+1}, y_t^{(n)}, x_t^{(m)}\right)}{p\left(y_t^{(n)}, x_t^{(m)}\right)} \qquad (1.35)$$

$$p\left(y_{t+1}|y_t^{(n)}\right) = \frac{p\left(y_{t+1}, y_t^{(n)}\right)}{p\left(y_t^{(n)}\right)} \qquad (1.36)$$

The file TRANS_ENT.CPP on my web site computes transfer entropy. It differs from the presentation just shown in one small way. The mathematical presentation uses the current and prior values of X and Y to predict the next value of Y to conform to already published work. But in programming terms, it is easier to use strictly historical values of X and Y to predict the current value of Y. These two approaches are equivalent, differing only in subscripts.

There is one feature in the program that adds versatility but is not represented in the mathematical presentation given earlier. So to make sure everything is clear, here is a rigorous statement of the problem addressed by the program:

- *y*: The series being predicted

- *x*: The series whose causative nature is being evaluated

- *n*: The length of each series

- *nbins_y*: The number of values that y can take on

- *nbins_x*: The number of values that x can take on

- *yhist*: The number of historic y observations used for prediction

- *xhist*: The number of historic x observations used for prediction

- *xlag*: See the problem statement and the comment that follows

We are given two series, x and y, each having n cases. It is assumed that p(y[i]) is a function of y[i-1], y[i-2], ..., y[i-yhist]. But does x[i-xlag], x[i-xlag-1], ..., x[i-xlag-xhist+1] influence the conditional state probabilities of y? This function measures the extent to which this occurs.

The traditional version of transfer entropy computation has xlag=1, meaning that the value of x concurrent with y is not allowed to participate in influencing y. However, many applications employ a dataset in which the X series is already implicitly lagged with respect to Y. For example, most model-based market-trading datasets compute X based strictly on the current and prior values of the market, and they compute Y based strictly on future values of the market. Rather than requiring the user to shift the data series or adjust addressing, this routine lets the user set xlag=0 to account for X already being lagged.

Note that we have nbins_x ^ xhist * nbins_y ^ (yhist+1) cells in the probability matrix corresponding to $(y_{t+1}, y_t^{(yhist)}, x_t^{(xhist)})$. (The symbol ^ means "raised to the power.") This blows up very, very quickly. For this reason, the majority of applications will use xhist=yhist=1 and have both nbins_x and nibins_y at most three, and often just two.

To clarify the program code, we use three single letters to represent the otherwise complex terms in the algorithm.

- *a*: The current value of y, which is being predicted
- *b*: The yhist historic values of y
- *c*: The xhist historic values of x

Using this compact notation, the transfer entropy of Equation (1.34) is expressed in the much less fierce Equation (1.37). Corresponding to Equations (1.35) and (1.36) we have p(a|b,c) = p(a,b,c) / p(b,c) and p(a|b) = p(a,b) / p(b).

$$Transfer\ entropy = \sum p(a,b,c) \log\left(\frac{p(a|b,c)}{p(a|b)}\right) \tag{1.37}$$

Now that this simpler notation is in place, we can present the routine in segments. It is called as shown here. Note that the values in x and y range from zero through nbins_x-1 and nbins_y-1, respectively.

```
double trans_ent (
  int n,          // Length of x and y
  int nbins_x,    // Number of x bins.
  int nbins_y,    // Ditto y
  short int *x,   // Independent variable, which impacts y transitions
  short int *y,   // Dependent variable
  int xlag,       // Lag of most recent predictive x: 1 for traditional, 0 for concurrent
  int xhist,      // Length of x history. At least 1
  int yhist,      // Ditto y
  int *counts,    // Work vector (see comments in code for length)
  double *ab,     // Ditto
  double *bc,     // Ditto
  double *b       // Ditto
  )
```

The first step is to compute several frequently used constants: nx=nbins_x^xhist and ny=nbins_y^yhist. This is done as follows:

```
nx = nbins_x;
for (i=1; i<xhist; i++)     // Number of bins for X history
  nx *= nbins_x;

ny = nbins_y;
for (i=1; i<yhist; i++)     // Number of bins for Y history
  ny *= nbins_y;

nxy = nx * ny;              // Total number of history bins
```

Count the number of cases that lie in each of the possible bins determined by the X history, the Y history, and the current value of Y. The counts are kept in an array with X history changing fastest, then Y history, and current Y changing last. We make sure not to start so early in the array that a negative subscript would be used.

```
memset (counts, 0, nxy * nbins_y * sizeof(int));
istart = xhist + xlag - 1;
if (yhist > istart)
  istart = yhist;

for (i=istart; i<n; i++) {

  // Which of the nbins_x ^ xhist X history bins does this case lie in?
  ix = x[i-xlag];
  for (j=1; j<xhist; j++)
    ix = nbins_x * ix + x[i-j-xlag];

  // Which of the nbins_y ^ yhist Y history bins does this case lie in?
  iy = y[i-1];
  for (j=2; j<=yhist; j++)
    iy = nbins_y * iy + y[i-j];

  ++counts [ y[i] * nxy + iy * nx + ix ]; // Increment the correct bin
  }

total = n - istart;
```

The next step is to compute the marginal probabilities, which will be used in later computation. This is just basic summation.

```
for (i=0; i<nbins_y*ny; i++)
   ab[i] = 0.0;
for (i=0; i<nx*ny; i++)
   bc[i] = 0.0;
for (i=0; i<ny; i++)
   b[i] = 0.0;

for (ia=0; ia<nbins_y; ia++) {
   for (iy=0; iy<ny; iy++) {
      for (ix=0; ix<nx; ix++) {
         p = (double) counts [ia * nxy + iy * nx + ix] / (double) total;
         ab[ia*ny+iy] += p;
         bc[iy*nx+ix] += p;
         b[iy] += p;
         }
      }
   }
```

Finally, we compute the transfer entropy. This is just a straightforward implementation of the defining equations.

```
trans = 0.0;
for (ia=0; ia<nbins_y; ia++) {
   for (iy=0; iy<ny; iy++) {
      for (ix=0; ix<nx; ix++) {
         p = (double) counts [ia * nxy + iy * nx + ix] / (double) total;     // p(a,b,c)
         if (p <= 0.0)
            continue;
         numer = p / bc[iy*nx+ix];        // p(a | b,c)
         denom = ab[ia*ny+iy] / b[iy];    // p(a | b)
         trans += p * log (numer / denom);   // Equation (1.37)
         }
      }
   }
```

We close this section by noting that my web site contains a program called TRANSFER.CPP (in the code set for my *Assessing...* book) that uses transfer entropy to sort a list of predictor candidates. This is similar to the SCREEN_UNIVAR.CPP program, so we will not bother listing it here. However, we will note one crucial difference between the two programs. SCREEN_UNIVAR.CPP shuffles the dependent variable to do the Monte Carlo permutations. This is the efficient way to do it, as there is only one dependent variable, while there are many independent candidates. But when data for transfer entropy is shuffled, we cannot take this approach. The reason is that shuffling the dependent variable would destroy any predictive power associated with its *own* historical values, when all we want to destroy is the relationship with the independent variable. Therefore, we must shuffle each candidate. Examination of the code will make clear how this is done.

CHAPTER 2

Screening for Relationships

Data miners are usually confronted with a daunting array of variables from which they hope to discover useful relationships. One could always just test them individually, in groups, or in a stepwise procedure, using a sophisticated model similar or identical to that which the developer wants to ultimately deploy. This direct approach would usually be the best in the sense that it would discover the relationships that will ultimately be most useful.

Unfortunately, in most situations, this direct approach is much too costly in terms of computational resources. Training sophisticated models can be horrendously slow and hence must be done with as little exploratory work as possible. Data miners need relatively fast screening procedures that can reduce a mountain of contenders to a much smaller subset of variables that are most likely to be useful in the application. This is the subject of this chapter.

Simple Screening Methods

Naturally, there are infinite methods for quickly screening candidate variables for relationships with one or more other variables (called the *target* variable or set of variables). However, a few are especially popular, and for good reasons. Thus, we will focus our in-depth presentation on those that are most commonly used, while lightly covering a few more that are uncommon but appropriate in special circumstances. Also note that relationships other than with regard to a target are possible. Some of these will be presented in the next chapter.

© Timothy Masters 2018
T. Masters, *Data Mining Algorithms in C++*, https://doi.org/10.1007/978-1-4842-3315-3_2

Univariate Screening

The most basic screening technique is to examine each candidate individually, looking at its relationship with the target without regard to any possible fortuitous interaction with other candidates. This method has the great advantage that it is fast, almost certainly the fastest of any of the common methods. This makes it mandatory whenever the developer has to deal with an unusually large number of candidates. But it does suffer from failing to make use of potentially vital interaction information. The classic example is predicting health risks from height and weight; the two together provide vastly more information than either alone.

Bivariate Screening

We can significantly alleviate the weakness of univariate screening by examining all possible *pairs* of candidates. This still does not allow us to capitalize on valuable interactions with a third variable, but in practice the information gain from taking candidates two at a time can be huge. Unfortunately, the cost can be prohibitive. For example, with 100 candidates there will be 100*99/2=4950 pairs to check. With 1,000 there will be almost half a million pairs. Unless the relationship criterion being evaluated is very fast to compute (such as with massive parallel processing), bivariate screening will be impractical when there are a large number of candidates.

Forward Stepwise Selection

This venerable algorithm has been in use for centuries (or at least it seems so). The idea is almost trivial. We find the single candidate variable that has the greatest relationship with the target. Then we find the variable that, if considered in conjunction with the one chosen first, adds the most to the relationship. Then we find a third variable from among the remaining candidates, which when considered in conjunction with the first two produces the greatest relationship with the target. This continues for as long as the developer desires.

The advantage of this method is that at each stage the number of candidate variables being tested for a relationship with the target is the minimum possible, thus delaying the devastation of a combinatoric explosion. The disadvantage is that it can easily produce a suboptimal set of predictors. For example, suppose X1 and X2 alone have little or no relationship with the target but together have a great relationship. And suppose X3 is modestly related to the target. If the user requests that two candidates be selected, X3 will be chosen first, and the wonderful X1, X2 pair will be missed. Never underestimate this issue; it can be devastating.

Forward Selection Preserving Subsets

There is a straightforward extension of forward stepwise selection that can often produce a significant improvement in performance at little cost. We simply preserve the best few candidates at each step, rather than preserving just the single best. For example, we may find that X4, X7, and X9 are the three best single variables. (Three is an arbitrary choice made by the developer, considering the trade-off between quality and compute time.) We then test X4 paired with each remaining candidate, X7 paired with each, and finally X9 paired with each. Of these many pairs tested, we identify the best three pairs. These pairs will each be tested with the remaining candidates as trios, and so forth. The beauty of this algorithm is that we gain a lot with relatively little cost. The chance of missing an important combination is greatly reduced, while compute time goes up linearly, not exponentially. I highly recommend this approach.

Backward Stepwise Selection

In the rare instance that computational resources allow, backward stepwise selection is optimal or close to it. The idea is that we throw *all* competitors into the pot and evaluate this group's relationship with the target. Then we find the single competitor whose elimination produces the *least* reduction in the relationship criterion. Keep eliminating this way until the remaining candidate set is the size desired by the developer.

Obviously, this method is only rarely practical. If the number of candidates is even moderately large, computation of the relationship criterion will almost certainly be impossible because of time constraints, accuracy (numerical stability) issues, memory requirements, or all of the above. Still, if you can pull it off, it usually doesn't get any better.

Criteria for a Relationship

Later in this chapter we will explore detailed algorithms that screen variables for relationships. But first, I present some of the most common and effective criteria for measuring the degree of a relationship between two variables. This will be extended to relationships between groups of variables in later sections.

Ordinary Correlation

Perhaps the oldest and most venerable measure of the relationship between two variables is *Pearson r*, often called just *correlation* (despite the fact that numerous alternative measures of correlation exist). It is sensitive to a *linear* relationship between them. Any curvature in their relationship will reduce their correlation, even if the actual relationship is strong. And if while one variable steadily increases but the other increases for a while and then decreases, we may find that their correlation is tiny, regardless of how strong their true relationship is. This can be a serious disadvantage. Another problem is that ordinary correlation is terribly sensitive to outliers (data values far outside the majority of values). Outliers will dominate the calculation, likely obscuring any legitimate relationship that exists within the mass of cases. Still, correlation is fast to compute, and it does capture many of the most common types of relationship. Thus, it is a vital member of our tool set.

Correlation ranges from −1, for a perfect inverse linear relationship, to +1 for a perfect positive linear relationship. A correlation of zero means that no linear relationship exists. If we have n pairs of values, x_i and y_i for i from 1 to n, then we compute the mean of x using Equation (2.1), and the mean of y similarly, and then compute their correlation with Equation (2.2).

$$\bar{x} = \frac{1}{n}\sum_{i=1}^{n} x_i \tag{2.1}$$

$$r = \frac{\sum_{i=1}^{n}(x_i - \bar{x})(y_i - \bar{y})}{\sqrt{\sum_{i=1}^{n}(x_i - \bar{x})^2 \sum_{i=1}^{n}(y_i - \bar{y})^2}} \tag{2.2}$$

Here is code for ordinary correlation, extracted from the file SCREEN_UNIVAR.CPP. It is a straightforward implementation of the prior equations.

```
static double compute_r (
   int ncases,          // Number of cases (rows) in data matrix
   int varnum,          // Column of predictor in database
   int n_vars,          // Number of columns in database
   double *data,        // The data is here; ncases rows by n_vars columns
   double *target       // The target (ncases long)
   )
```

```
{
  int icase;
  double xdiff, ydiff, xmean, ymean, xvar, yvar, xy;

  xmean = ymean = 0.0;
  for (icase=0; icase<ncases; icase++) {      // Equation (2.1)
    xmean += data[icase*n_vars+varnum];    // Get predictor candidate 'varnum'
    ymean += target[icase];                // The target is separate from candidates
  }
  xmean /= ncases;
  ymean /= ncases;

  xvar = yvar = xy = 1.e-30;                   // Prevent division by zero later
  for (icase=0; icase<ncases; icase++) {      // Equation (2.2)
    xdiff = data[icase*n_vars+varnum] - xmean;
    ydiff = target[icase] - ymean;
    xvar += xdiff * xdiff;
    yvar += ydiff * ydiff;
    xy += xdiff * ydiff;
  }

  return xy / sqrt (xvar * yvar);
}
```

Nonparametric Correlation

A serious problem with ordinary correlation (Pearson r) is its sensitivity to outlying data values. Even one wild data point can render ordinary correlation worthless. This can be remedied by ranking each of the two variables from smallest to largest and determining the degree to which their ranks correspond (small ranks of one variable correspond to small ranks of the other, and similarly for large ranks). A common and highly effective rank-based measure of correlation is *Spearman rho*. Suppose we recompute the two variables, assigning to each a value of 1 for the smallest value of that variable, 2 for the second-smallest, and so forth. Subsequent calculations are based on these ranks rather than the raw data.

If either variable has tied values, we must compensate for these ties. For each tied value, assign to all members of the tied set the mean rank that they would have if they were not tied. Let $t_{k,X}$ be the number of tied values at a given rank for the X variable. Let $TieCorrection_{k,X}$ be given by Equation (2.3). Let $SumTieCorrection_X$ be the sum of $TieCorrection_{k,X}$ for the X variable, as expressed in Equation (2.4). Define SS_X as shown in Equation (2.5). These quantities are defined similarly for the Y variable. For each case, compute the difference in ranks, and sum these squared differences, as shown in Equation (2.6). Remember that in this equation, the x and y values refer to ranks, not the original data. Finally, compute the Spearman rho with Equation (2.7). The code for computing Spearman rho, extracted from SCREEN_UNIVAR.CPP, follows these equations.

$$TieCorrection_{k,X} = t_{k,X}^3 - t_{k,X} \tag{2.3}$$

$$SumTieCorrection_X = \sum_k TieCorrection_{k,X} \tag{2.4}$$

$$SS_X = \frac{1}{12}\left(n^3 - n - SumTieCorrection_X\right) \tag{2.5}$$

$$D = \sum_{i=1}^{n}\left(x_i - y_i\right)^2 \tag{2.6}$$

$$\rho = \frac{SS_X + SS_Y - D}{2\sqrt{SS_X SS_Y}} \tag{2.7}$$

```
static double compute_rho (// Returns Spearman rho in range -1 to 1
   int ncases,       // Number of cases (rows) in data matrix
   int varnum,       // Column of predictor in database
   int n_vars,       // Number of columns in database
   double *data,     // The data is here; ncases rows by n_vars columns
   double *target,   // The target (ncases long)
   double *x,        // Work vector ncases long
   double *y         // Work vector ncases long
   )
{
   int icase, j, k, ntied;
   double val, x_tie_correc, y_tie_correc, dn, ssx, ssy, rank, diff, rankerr, rho;
```

```
// We need to rearrange input vectors, so copy them to work vectors
for (icase=0; icase<ncases; icase++) {
   x[icase]= data[icase*n_vars+varnum];   // Fetch predictor 'varnum' from database
   y[icase] = target[icase];              // The target is kept separate
   }

// Compute ties in x, compute correction as SUM (ties**3 - ties)
// The following routine sorts x ascending and simultaneously moves y
qsortds (0, ncases-1, x, y);

x_tie_correc = 0.0;
for (j=0; j<ncases;) {                 // Convert x to ranks, cumulate tie corec
   val = x[j];                         // X for this case
   for (k=j+1; k<ncases; k++) {        // Find all ties
     if (x[k] > val)
       break;
     }
   ntied = k - j;                      // t_{k,X}
   x_tie_correc += (double) ntied * ntied * ntied - ntied; // Equations (2.3) and (2.4)
   rank = 0.5 * ((double) j + (double) k + 1.0);   // Tied rank is mean rank across ties
   while (j < k)                               // Assign this value to all ties here
     x[j++] = rank;
   } // For each case in sorted x array

// Now do same for y
qsortds (0, ncases-1, y, x);
y_tie_correc = 0.0;
for (j=0; j<ncases;) { // Convert y to ranks, cumulate tie corec
   val = y[j];
   for (k=j+1; k<ncases; k++) {   // Find all ties
     if (y[k] > val)
       break;
     }
   ntied = k - j;
   y_tie_correc += (double) ntied * ntied * ntied - ntied; // Equations (2.3) and (2.4)
   rank = 0.5 * ((double) j + (double) k + 1.0);      // Tied rank is mean rank across ties
```

```
    while (j < k)                            // Assign this value to all ties here
        y[j++] = rank;
    } // For each case in sorted y array

    // Final computations
    dn = ncases;
    ssx = (dn * dn * dn - dn - x_tie_correc) / 12.0;    // Equation (2.5)
    ssy = (dn * dn * dn - dn - y_tie_correc) / 12.0;

    rankerr = 0.0;
    for (j=0; j<ncases; j++) {      // Cumulate squared rank differences
        diff = x[j] - y[j];
        rankerr += diff * diff;       // Equation (2.6)
    }

    rho = 0.5 * (ssx + ssy - rankerr) / sqrt (ssx * ssy + 1.e-20); // Equation (2.7)
    return rho;
}
```

Accommodating Simple Nonlinearity

Ordinary correlation and Spearman rho respond to linear relationships between variables, while many real-life variables have nonlinear relationships that are difficult to quantify with these measures. Later in this chapter we will explore powerful general-purpose information-based algorithms for discovering any relationship between variables, even if the relationship is profoundly nonlinear. But those methods can have drawbacks of their own, such as excessive runtime, troublesome sensitivity to user-specified parameters, and suboptimal exploitation of observed values of variables. There is a middle ground that can be useful in many applications.

The concept is simple: designate one variable as a "target" to be predicted and the other variable as a predictor. Compute a least-squares quadratic equation for predicting the target from the predictor. Then the measure of relationship is the R-squared of this prediction.

The advantages of this method are similar to those of ordinary correlation: it is relatively fast to compute, it does not require that the user specify any parameters, and it makes excellent use of all information contained in the variables. Nonetheless, it responds not only to a linear relationship but also to the sort of curvature often found in real-life variables, going so far as to even handle complete reversal of the relationship across the range. This is a powerful property.

It is worth noting (though usually of little practical consequence) that unlike most other criteria described in this section, this method is not symmetric. Reversing the roles of the predictor and the target variable will produce different results. But in most applications, directionality is inherent, so the labeling is natural.

I will not go into the mathematical derivation of this least-squares fit. It is tedious and well covered in numerous other sources. However, I will present the source code and point out that the fit is done with singular value decomposition. See the file SVDCMP.CPP for more details on this excellent fitting method. The criterion computation code, extracted from SCREEN_UNIVAR.CPP, is shown here:

```
static double compute_quad (
   SingularValueDecomp *sptr, // Used for finding optimal coefficients
   int ncases,          // Number of cases (rows) in data matrix
   int varnum,          // Column of predictor in database
   int n_vars,          // Number of columns in database
   double *data,        // The data is here; ncases rows by n_vars columns
   double *target       // The target (ncases long)
   )
{
   int icase;
   double xdiff, ydiff, xmean, ymean, xstd, ystd;
   double *aptr, *bptr, coefs[3], sum, mse;
/*
   Standardize the data for stability and simplified calculation.
   Making the target have unit variance means that the mse is the unpredictable fraction.
   Making the predictors have smallish mean and similar variance helps stability.
*/

   xmean = ymean = 0.0;
   for (icase=0; icase<ncases; icase++) {
      xmean += data[icase*n_vars+varnum];   // Get this predictor variable
      ymean += target[icase];               // The target is kept separate
      }
```

```
xmean /= ncases;
ymean /= ncases;

xstd = ystd = 1.e-30;
for (icase=0; icase<ncases; icase++) {
   xdiff = data[icase*n_vars+varnum] - xmean;
   ydiff = target[icase] - ymean;
   xstd += xdiff * xdiff;
   ystd += ydiff * ydiff;
   }

xstd = sqrt (xstd / ncases);
ystd = sqrt (ystd / ncases);

/*
   Fill in SingularValueDecomp object and compute optimal coefficients
*/

aptr = sptr->a;
bptr = sptr->b;

for (icase=0; icase<ncases; icase++) {
   xdiff = (data[icase*n_vars+varnum] - xmean) / xstd;
   ydiff = (target[icase] - ymean) / ystd;
   *aptr++ = xdiff * xdiff; // Quadratic term
   *aptr++ = xdiff;         // Linear term
   *aptr++ = 1.0;           // Constant term
   *bptr++ = ydiff;         // Predicted value
   }

sptr->svdcmp ();
sptr->backsub (1.e-7, coefs);

/*
   Compute the error. We pass through the data. For each case, predict the target and sum the mean
   squared error of the prediction.
*/

mse = 0.0;
```

```
for (icase=0; icase<ncases; icase++) {
    xdiff = (data[icase*n_vars+varnum] - xmean) / xstd;      // Standardized predictor
    ydiff = (target[icase] - ymean) / ystd;                  // Standardized target
    sum = coefs[0] * xdiff * xdiff + coefs[1] * xdiff + coefs[2]; // Prediction
    ydiff -= sum;                    // True minus predicted is error of this prediction
    mse += ydiff * ydiff;           // Cumulate mean squared error
    }

    return 1.0 - mse /ncases;       // Target is standardized, so this is R-squared
}
```

It should be noted that when the SingularValueDecomp object is created, we could specify that the *a* matrix be preserved for reuse in the error computation. This avoids repeating the standardization, at the cost of more memory. The choice is yours.

Chi-Square and Cramer's V

When two variables are categorical (gender, college major, political affiliation, etc.), the standard way to assess their degree of relationship is the chi-square test. We create a matrix in which the categories of one variable form the rows, and those of the other form the columns. The occurrences of each possible pairing of categories are counted within the dataset being analyzed. The expected count for each pairing under the assumption that the variables are unrelated is computed and then compared to the actually observed counts. The more the observed counts depart from their expected values, the more the variables are related.

But the chi-square test need not be restricted to categorical variables. It is legitimate to partition the range of numeric variables into bins and treat these bins as if they were categories. Of course, this results in some loss of information because variation within each bin is ignored. But if the data is noisy or if one wants to detect relationship patterns of any form without preconceptions, a chi-square formulation may be appropriate.

Suppose variable X is partitioned into K_X bins, and variable Y is partitioned into K_Y bins. Let $N_{X,i}$ be the number of cases whose variable X falls in bin i. Similarly, let $N_{Y,j}$ be the number of cases whose variable Y falls in bin j. The total number of cases is N. Then the *marginal distribution* of X is given by Equation (2.8), and similarly for Y.

$$F_X(i) = \frac{N_{X,i}}{N} \tag{2.8}$$

Suppose X and Y are unrelated. The probability that a case will be in bin i for X and bin j for Y is the product of the marginals, as shown in Equation (2.9). The expected number of cases in this conjunction of bins is this probability times the total number of cases, as shown in Equation (2.10).

$$F_{X,Y}(i,j) = F_X(i)F_Y(j) \tag{2.9}$$

$$E_{i,j} = N F_{X,Y}(i,j) \tag{2.10}$$

Let $O_{i,j}$ be the observed number of cases in bin i for X and bin j for Y. If X and Y are unrelated, this quantity will tend to be close to $E_{i,j}$, the expected cell count under the assumption of independence. But if the variables are related, then some combinations of bins will be favored, while others will be less common. This departure from expectation is computed with Equation (2.11).

$$ChiSquared = \sum_i \sum_j \frac{(O_{i,j} - E_{i,j})^2}{E_{i,j}} \tag{2.11}$$

Chi-squared itself has little intuitive meaning in terms of its values. It is highly dependent on the number of cases and the number of bins for each variable, so any numeric value of chi-squared is essentially uninterpretable. This can be remedied by a simple monotonic transformation to produce a quantity called *Cramer's V* shown in Equation (2.12). This ranges from zero (no relationship between X and Y) to one (perfect relationship).

$$V = \sqrt{\frac{\frac{ChiSquare}{N}}{\min(K_X - 1, K_Y - 1)}} \tag{2.12}$$

Here is code for computing Cramer's V. This is extracted from the file SCREEN_UNIVAR.CPP. The calling parameter list is as shown here. The routine follows. The marginals, shown in Equation (2.8), are already computed prior to calling this routine to save redundant effort.

```
static double compute_V (
   int ncases,              // Number of cases
   int nbins_pred,          // Number of predictor bins
   int *pred_bin,           // Ncases vector of predictor bin indices
```

```
int nbins_target,          // Number of target bins
int *target_bin,           // Ncases vector of target bin indices
double *pred_marginal,     // Predictor marginal
double *target_marginal,   // Target marginal
int *bin_counts            // Work area nbins_pred*nbins_target long
)

{
  int i, j;
  double diff, expected, chisq, V;

  for (i=0; i<nbins_pred; i++) {   // Zero bin counts
    for (j=0; j<nbins_target; j++)
      bin_counts[i*nbins_target+j] = 0;
  }

  for (i=0; i<ncases; i++)      // Cumulate bin counts Oij
    ++bin_counts[pred_bin[i]*nbins_target+target_bin[i]];

  chisq = 0.0;
  for (i=0; i<nbins_pred; i++) {
    for (j=0; j<nbins_target; j++) {
      expected = pred_marginal[i] * target_marginal[j] * ncases; //Equation (2.9), (2.10)
      diff = bin_counts[i*nbins_target+j] - expected;
      chisq += diff * diff / (expected + 1.e-20); // Equation (2.11)
    }
  }

  V = chisq / ncases;  // This and remaining lines are Equation (2.12)
  if (nbins_pred < nbins_target)
    V /= nbins_pred - 1;
  else
    V /= nbins_target - 1;

  V = sqrt (V);

  return V;
}
```

Mutual Information and Uncertainty Reduction

Mutual information and uncertainty reduction were thoroughly discussed in the prior chapter, so I will gloss over them quickly here, reviewing them only in the context of univariate screening.

These two measures of association are similar to the chi-square/Cramer's V measures of the prior section in that they rely on partitioning the range of the variables into discrete bins (although we did see a way of computing mutual information from continuous data). In fact, in many applications, the chi-square method and the mutual information method will give similar results. The actual numbers will be different, of course, but the ordering of a set of candidate predictors will often be almost identical. Nonetheless, they do measure slightly different quantities, so it is in our best interest to include both in our toolbox.

I should also remind you that uncertainty reduction is asymmetric; one variable must be designated as a predictor, and the other as a target (predicted). Reversing this labeling will produce different results. This is usually a good property because most applications have this same asymmetry.

Multivariate Extensions

The chi-square and information-based measures have been presented in the context of quantifying the relationship between two variables. However, it is easy to extend them to multiple variables. There are two completely different approaches to this.

The first and most common approach assumes we want to measure the degree to which one or more variables, taken as a set, are related to one or more other variables, also taken as a set. There is just one relationship we are interested in, although one or both sides of this relationship may be a set of variables rather than just a single variable. I'll present a useful application of this on page 116.

The method is simple: just unwrap the bins in each set, producing a new set of bins on each side whose dimension is the product of the number of bins in the unwrapped side. For example, suppose we are assessing the relationship between X and Y, considered together, with Z. Suppose we have divided X into 2 bins, Y into 3, and Z into 4. We unwrap X and Y into 6 bins, one for each of the 2*3 possible combinations of X and Y. This gives us a 6-by-4 matrix on which we can perform our usual chi-square or information-based calculations.

Another multivariate extension, not often used, allows the user to test for a group relationship, an association among more than two variables. In this case, we create a three-dimensional (or however many variables are tested) grid. Equation (2.8) is used to compute the marginal across each dimension; Equation (2.9) gives the individual cell probabilities, extended to higher dimensions as needed; Equation (2.10) gives the expected cell counts; and Equation (2.11), extended to the requisite number of dimensions, gives the chi-square value. However, traditional probability calculations and a conversion to Cramer's V no longer apply in this case. We must use Monte Carlo permutation tests (described in the next chapter) to evaluate the significance of results.

Permutation Tests

Many of the measures of association described in prior sections have sufficient theoretical understanding among experts that we could use traditional exact statistical tests to compute the probability that an observed strong relationship could have arisen from luck alone, with the variables in fact being unrelated. However, not all of these measures have this property. Also, some of the tests (such as for chi-square with small cell counts) are far from accurate. But most importantly, when we are data mining, luck plays a disturbingly large role if we search for relationships among a large number of candidate variables. Thus, traditional statistical tests usually take a back seat to specialized tests aimed at dealing with the various ways that random luck can invalidate apparently correct results. In this section, we will examine a family of such tests that is extremely powerful and useful in data mining.

A Modestly Rigorous Statement of the Procedure

We begin with some potentially intimidating mathematics behind the tests to be soon described. Be assured that you can safely skip this section. But for those who care…

Suppose we have a scalar-valued function of a vector. We'll call this $g(v)$. In our current context, v would be the vector of cases for one variable (typically the target, if one is using such a label) and $g(.)$ would be a measure of association of this variable with another variable (typically a predictor candidate). This might be any of the measures described in the prior section.

Let $\Phi(.)$ be a permutation. In other words, $\Phi(v)$ is the vector v rearranged to a different order. Suppose v has n elements. Then there are $n!$ possible permutations. We can index these as Φ_i where i ranges from 1 through $n!$. For the moment, assume that the function value of every permutation is different: $g(\Phi_i(v)) \neq g(\Phi_j(v))$ when $i \neq j$. We'll discuss ties later.

Define $\Phi^\star$ as the *original* permutation, the ordering of v that is observed in the experiment and that corresponds to the order of the other variable. This is the arrangement of pairings that the universe happened to provide in our real life. Now draw from the population of possible orderings m more times and similarly define Φ_1 through Φ_m. Again, for the moment, assume that we force these $m+1$ draws to be unique, perhaps by doing the draws without replacement. We'll handle ties later.

Compute $g(\Phi^\star(v))$ and $g(\Phi_i(v))$ for i from 1 through m. Define the statistic Θ as the fraction of these $m+1$ values that are less than or equal to $g(\Phi^\star(v))$. Suppose the distribution of $g(\Phi(v))$ under sampling of v from the universe of possible values for this variable does not depend on Φ. This is the null hypothesis. In the current context, this means that among the population of possible values of the target, the distribution of our relationship measure does not depend on the ordering of the observed values of the target; the target and the predictor have no relationship Then the distribution of Θ does not depend on the labeling of the permutations, or on $g(.)$. In fact, Θ follows a uniform distribution over the values $1/(m+1)$, $2/(m+1)$, ..., 1. This is easy to see. Sort these $m+1$ values in increasing order. Because each of the draws that index the permutations has equal probability and because we are (temporarily) assuming that there will be no ties, the order is unique. Therefore, $g(\Phi^\star(v))$ may occupy any of the $m+1$ ordered positions with equal probability.

Let $F(\Theta)$ be the cumulative distribution function of Θ. As m increases, $F(\Theta)$ converges to a continuous uniform distribution on $(0,1)$. In other words, the probability that Θ will be less than or equal to, say, 0.05 will equal 0.05, and the probability that Θ will exceed, say, 0.99 will be 0.01, and so forth.

We can use this fact to define a statistical test of the null hypothesis that $\Phi^\star$, our original permutation, is indeed a random draw from among the $n!$ possible permutations, as opposed to being a special permutation that has an unusually large (or small) value of $g(\Phi^\star(v))$, the measure of relationship. To perform a left-tail test (unusually small relationship), set a threshold equal to the desired p-value, and reject the null hypothesis if the observed Θ is less than or equal to the threshold. To perform a right-tail test (unusually large relationship), set a threshold equal to one minus the desired p-value, and reject the null hypothesis if the observed Θ is greater than the threshold.

We have conveniently assumed that every permutation gives rise to a unique function value and that every randomly chosen permutation is unique. This precludes ties. However, the experimental situation may prevent us from avoiding tied function values, and selecting unique permutations is tedious. We are best off simply taking possible ties into account. Note that when comparing $g(\Phi^\star(v))$ to its m compatriots, tied values that are strictly above or below $g(\Phi^\star(v))$ are irrelevant. We only need to worry about ties at $g(\Phi^\star(v))$. A left-tail test will be conservative in this case. Unfortunately, a right-tail test will become anti-conservative. The solution is to shift the count boundary to the conservative end of the set of ties. The code shown later actually computes conservative p-values directly, and it slightly modifies the counting procedure accordingly.

Remember that an utterly crucial assumption for this test is that when the null hypothesis is true (the variables are unrelated), all of the $n!$ possible permutations, including of course the original one, have an equal chance of appearing, both in real life and in the process of randomly selecting m of them to perform the test. Violations of this assumption can creep into an application in subtle ways. The most common culprit, serial correlation in both variables, will be addressed later in this section.

A More Intuitive Approach

I suspect that most readers skipped over the theoretical discussion just shown. That's fine. Here is a more intuitive look at permutation tests.

The scenario under which this particular test might be employed is as follows: We have two variables, which for the sake of clarity we will call the *predictor* and the *target*, though they need not have this directional relationship. We choose a test statistic that will measure the relationship between these two variables. This may be mutual information, Cramer's V, or any other statistic that we favor. We then compute our measure of relationship.

A naive experimenter would look at the computed relationship figure and, if it is impressive, capitalize on this relationship in some way. But there is an aspect of the relationship measure that is every bit as important as its magnitude: the probability that truly unrelated variables could have scored as well by virtue of good luck. If this probability is anything but tiny, we must be skeptical.

Here is one way to handle this situation. Suppose we randomly permute one of the variables, typically the target. This destroys any actual relationship between the unpermuted predictor and the permuted target. They are now randomly paired up. We recompute the relationship measure. If this value is less than that obtained from the raw, unpermuted data, we are happy for this small bit of evidence that the two variables are truly related. But it's not very convincing evidence. If the variables were truly unrelated, there would still be a 50-50 chance of observing this result. So we need to test more random permutations.

If we test nine random permutations and the relationship measure for the original data exceeds all of them, we have more convincing evidence. In particular, if the variables were unrelated, there is a 1/10 chance that good luck would have placed it at the top. After all, in this situation, any of these ten orderings of the changes (one of them being the original order) has an equal shot at being the best.

What if the original relationship measure is the second best of the ten? There is a 2/10 probability that it will land in the best or second-best slot. So, suppose we had decided in advance that if the original measure is at least the second best, we would confidently conclude that our variables are related. If in truth they are unrelated, we would have a 20 percent chance of being fooled by good luck.

Suppose we decide in advance to conclude that our variables are related if the relationship measure on the original data has at least a specified rank among all permutations. It should be apparent that there is a simple formula for computing the probability of this event under the scenario that the variables are unrelated.

Let m be the number of random permutations tested (not counting the original), and let k be the number of these random permutations (again, not counting the original) whose relationship measure equals or exceeds that of the original. Then, the probability that the original measure will achieve this exalted position or better by sheer luck is $(k+1) / (m+1)$. You can understand this formula if you visualize the $m+1$ statistics (original plus m permuted) lined up in order. Note that the original statistic has equal probability of occupying any of these $m+1$ slots if the variables are unrelated.

A traditional statistical test of the null hypothesis that the variables are unrelated, versus the alternative that they are, would be performed as follows: Decide in advance what level of error probability you are willing to accept. This error, often called the *alpha level*, is an upper bound for the probability that you will erroneously reject the null hypothesis. Here, this error is concluding that the variables are related when in fact they are not. Choose a large value of m, and compute k from the previous formula. Then perform the random replications and count how many of them have a relationship statistic that equals or exceeds that of the original data. If k of them or fewer do so, we can reject the null hypothesis. If the null hypothesis is true (the variables are unrelated), we will make this error with probability at most our specified alpha.

Serial Correlation Can Be Deadly

Recall that a fundamental assumption of a Monte Carlo permutation test is that every possible permutation must be equally likely if the null hypothesis is true. If there is any sort of dependence in the vector being permuted, with serial correlation being by far the most common, then full permutation will destroy this serial correlation. This makes the test anti-conservative, more likely to indicate that a relationship is present when it is not. This is an extremely serious error.

But note that this is a problem only if *both* vectors contain dependencies. As long as at most one of the two variables has dependencies, we can permute the other one. And if we are using a symmetric measure of relationship, we can even permute the dependent variable because this revised pairing is equivalent to permuting the "good" variable!

In the next section, we will see a permutation algorithm that does a good (though not perfect) job of handling the situation of both variables having serial correlation.

It must be emphasized that this phenomenon is not an artifact of just the Monte Carlo permutation test. This is a universal phenomenon, which is why Statistics 101 courses always emphasize the importance of independent observations. The simple explanation of why this occurs is that any sort of dependence reduces the effective degrees of freedom of the test. The testing procedure looks at the number of cases and proceeds accordingly, but the dependence in the data increases the variance of the test statistic beyond what would be expected from a sample of the given size. Thus, we are more likely to falsely reject the null hypothesis.

Permutation Algorithms

Surprising as it may seem, permutation can be a significant eater of time in a Monte Carlo permutation test. It is not unusual for permuting a variable to require about as much computer time as computing the relationship criterion. Therefore, we must program it as efficiently as possible, paying special attention to the speed of the random number generator. Here is the "standard" permutation algorithm:

```
i = n_cases;      // Number remaining to be shuffled
while (i > 1) {    // While at least 2 left to shuffle
  j = (int) (unifrand_fast () * i); // Random must range from 0 (inclusive) to 1 (exclusive)
  if (j >= i)       // This should not be necessary, but safety is good
    j = i - 1;
  dtemp = target[--i];   // Swap i and j cases
  target[i] = target[j];
  target[j] = dtemp;
  }
```

If both variables have serial correlation, there is an alternative shuffling algorithm that greatly reduces (though it does not completely eliminate) the deadly anti-conservative behavior of ordinary shuffling. Still, any anti-conservative tendency is scary, so we should exercise care in interpreting these results. But this algorithm is better than nothing and is perfectly reasonable for rough results.

The idea is that instead of swapping cases randomly, we rotate the permuted series. This keeps serial dependencies largely intact, but it still destroys the pairing of values of the two series and hence destroys the relationship between the series, which is what must do to generate the null hypothesis distribution. Here is this rotational permutation algorithm. Note that we use a scratch vector, work_target.

```
j = (int) (unifrand_fast () * n_cases);  // Rand ranges from 0 (inclusive) to 1 (exclusive)
if (j >= n_cases)                 // Should not be necessary, but play it safe
  j = n_cases - 1;
for (i=0; i<n_cases; i++)        // Rotate into work vector
  work_target[i] = target[(i+j)%n_cases];
for (i=0; i<n_cases; i++)        // Copy rotated vector back into target vector
  target[i] = work_target[i];
```

Outline of the Permutation Test Algorithm

Later, we will explore specific versions of the Monte Carlo permutation test, adapted for specialized applications. However, before advancing further, I will summarize the material shown so far by presenting a general outline of the most basic procedure. This will serve as a foundation for more sophisticated applications. Here it is in words:

```
for permutation from 0 through n_permutes-1

    if permutation > 0
        shuffle one variable (typically the target)

    compute 'criterion', the measure of relationship

    if permutation = 0
        original criterion = criterion
        count = 1

    else
        if criterion >= original criterion
            count = count + 1

probability = count / n_permutes
```

The probability computed by this algorithm is the approximate probability that, if the two variables are truly unrelated, a measure of their relationship at least as large as that observed could be obtained by pure good luck. If you find a wonderfully nice relationship, before trying to capitalize on it, you should run this test and confirm that the computed probability is small. If it is not small, you should be highly suspicious of your results. Undetected good luck has a way of coming back to bite you when you least expect it.

Just to dot all my i's and cross all my t's, I'll note that rejecting a potential relationship based on a nonsmall probability is perilously close to a sin that statisticians call *accepting a null hypothesis*, a serious no-no. Thus, we must avoid saying that a relationship with a nonsmall probability is worthless. We should just be suspicious, especially if the sample is large.

Permutation Testing for Selection Bias

We come now to what I believe is the most important use of Monte Carlo permutation tests: accounting for selection bias (the bias inherent in selecting the best of many competitors). The problem with the probability computed with the algorithm just shown is that if more than one predictor candidate is tested for a relationship with a target (the usual situation!), then there is a large probability that some truly worthless candidate will be lucky enough to achieve a high level of the relationship measure and hence achieve a very small probability. In fact, if all candidates are worthless, the probabilities of the candidates will follow a uniform distribution, frequently obtaining small values by random chance. This situation can be remedied by conducting a more advanced test that accounts for this selection bias. The *unbiased probability* for the *best* performer in the candidate set is the probability that this best performer could have attained its exalted level of performance by sheer luck if *all* candidates were truly worthless.

We can easily compute the unbiased probability for all candidates, not just the best. For those other, lesser candidates, the computed unbiased probability is an upper bound (a conservative measure) for the true unbiased probability of the candidate. Thus, a very small unbiased probability for any candidate is a strong indication that the candidate has true predictive power. Unfortunately, unlike the regular (often called the *solo*) probability, large values of the unbiased probability are not necessarily evidence that the candidate is worthless. Large values, especially near the bottom of the sorted list of relationship measures, may be due to over-estimation of the true unbiased probability. I am not aware of any algorithm for computing correct unbiased probabilities for any candidate other than the best. However, because this measure is conservative, it does have great utility in selecting promising predictors.

The algorithm, modified to handle selection bias, is shown here:

```
for permutation from 0 through n_permutes-1

    if permutation > 0
        shuffle the target

    for 'variable' covering all predictor candidates
        compute 'criterion', the measure of relationship between variable and target
        if permutation = 0
            original criterion[variable] = criterion
            solo_count[variable] = unbiased_count[variable] = 1
        else
            if criterion[variable] >= original criterion[variable]
                solo_count[variable] = solo_count[variable] + 1

    if permutation > 0
        best_criterion = MAX (criterion for all predictor candidates)
        for 'variable' covering all predictor candidates
            if best_criterion >= original_criterion[variable]
                unbiased_count[variable] = unbiased_count[variable] + 1

for 'variable' covering all predictor candidates
    solo_probability[variable] = solo_count[variable] / n_permutes
    unbiased_probability[variable] = unbiased_count[variable] / n_permutes
```

The first step to understanding this algorithm is to note that for the solo probabilities, for each candidate predictor this is identical to the simple algorithm shown on page 94.

But this algorithm contains one additional step. For shuffled runs, it finds the maximum of the relationship measures for all candidates. Then, for each candidate, it compares this "best" measure to the original score for the candidate and increments the unbiased counter accordingly. For whichever candidate has the greatest original relationship, this is in perfect conformation: the greatest measure for permuted data is compared to the greatest measure for the original data. Hence, this provides the probability that, if all candidates were worthless, the obtained best relationship could have been obtained by pure luck. But do note that for candidates other than the best, this probability is conservative.

Combinatorially Symmetric Cross Validation

The primary goal of most data mining operations is not just discovery of relationships that exist within a dataset that is in our hands. Rather, what we really want is to discover relationships that exist in the general population of interest. It does us little good (and perhaps great harm!) if we collect a dataset, analyze the daylights out of it, proudly proclaim a momentous discovery, and then learn that our discovery cannot be reproduced in subsequent data collections. Such a situation is usually associated with *overfitting* our relationship model.

We saw one approach to dealing with this issue in the prior section, when we used a permutation test to estimate the probability that results as good as those observed could have been obtained by pure luck. In this section, we take a completely different approach. It is based on the fact that the data in our sample contains two components: true values and random noise. For every variable measured in every case, the value in our dataset is composed of an unobservable true value plus contamination by noise. So when we measure the relationship between variables, we are not getting a measure of the relationship between the true values. Instead, we are measuring the relationship between our observed values, which for all we know may consist of more noise than truth! Especially if many variables are under investigation, it may be that a randomly fortuitous alignment of noise patterns may result in deceptive relationships that do not exist in the general population.

This is particularly problematic if our measure of relationship is overly powerful. To take an extreme example, a careless developer may postulate that a dependent variable is related to an independent variable by a degree-ten polynomial and measure the degree of relationship by the R-squared of the fit. In the vast majority of applications, this would be called *overfitting*, because the measure is much too capable of capitalizing on phantom relationships between the noise components. As a less extreme but still serious example, if we were to compute a bin-based measure such as discrete mutual information or Cramer's V and use a bin resolution that is too fine, we could find nonreproducible relationships between the noise components.

The *CSCV* algorithm presented in this section, which is loosely based on ideas given in "The Probability of Backtest Overfitting" by David Bailey, Jonathan Borwein, Marcos Lopez de Prado, and Jim Zhu, is much more context-sensitive than the Monte Carlo permutation testing of the prior section. The *theoretical* (though not necessarily practical) assumption is that, in some sense best left undefined, the set of variables competing in a relation contest with some other variable is complete and representative. Roughly speaking, this means that the tested competitors encompass all possible competitors in the application and do not include any variables that do not naturally fit in the application.

Okay, I know. Quit rolling your eyes. Not only is this description vague, but it is also impossible to achieve in real life. The good news is that, in practice, violations of this assumption, unless they are outrageously egregious, are almost always of little or no consequence. The main thing we need to be concerned with is that we do not include in the competition any variables that a reasonable person would know in advance have nothing to do with the application. Accidental inclusion of worthless variables is not a serious problem; in fact, this is usually impossible to avoid in practical data mining. Just don't deliberately include crazy things.

For example, suppose we are hoping to discover personal traits that predict the efficacy of some new drug. We would certainly include the person's age, weight, gender, blood type, and so forth. We might even stretch a little by including the person's hair color, hobbies, pets in their home, and other traits that have no obvious relationship to drug response. But we should not include the Dollar/Yen foreign exchange rate on the day they were born. Inclusion of too many such variables will distort results.

Also, we should not cheat by deliberately omitting competitors that we know in advance may have a reasonable chance of being useful. In the earlier drug example, we must not say, "I know from experience that weight will be a powerful predictor, so there is no sense even testing it." Such an omission will seriously distort results. Of course, if you accidentally omit a useful predictor, so be it. You can't always know in advance everything that is useful. Just don't do it deliberately.

Let's pause for a moment and digress into the fact that the CSCV algorithm is far more general than its presentation here. In this text and subsequent code, we employ it for one purpose, as an aid for evaluating relationships between individual competing variables and a single other variable. On page 102 we will see the algorithm in its most general version, and at that point it should be clear how to generalize it. Here are a few examples of how CSCV can aid in the evaluation of competing multiple comparisons:

- One group of variables is jointly related to another group of variables. Choose the variables that make up each set so as to maximize their joint relationship.

- A model has numerous competing sets of parameters. In other words, the competitors are parameter values rather than variables, and we find the most effective parameter set.

- A financial market trading system has competing versions or parameter sets. This is the application that [Bailey et al, 2015] considers.

Now that the preliminaries are out of the way, let's talk about exactly what we will be doing in this test. We have collected a sample of data, our dataset, and computed performance statistics for the competitors. Because our performance statistics are based on a sample that is contaminated by noise, our computed values will not exactly equal the (unmeasurable) true values in the population from which our sample was drawn. We hope that they are close. In particular, when we determine the best competitor, that having the maximum performance statistic, we hope that its true performance in the population is also outstanding.

What is a good criterion to use in order to define "outstanding" performance out-of-sample (not in our dataset)? The choice employed for this test is to compare the out-of-sample (*OOS*) performance of the best competitor (or any competitor in general) to the *median* OOS performance of all competitors. It's a fairly low bar, but we define outstanding performance as being above the median. If a competitor's OOS performance is above the median OOS performance of all competitors, we say that this competitor is outstanding.

Now it should be clear why the field of competitors should be "complete" and "representative" for the application. Suppose some competitors that are known a priori to be useful are omitted. The median will be skewed downward from what it would be in a fair fight. Similarly, suppose we include a bunch of competitors that a reasonable person would know in advance to be useless. In this case we have again deliberately skewed the median downward. In either case, the relative performance of our competitors will be inflated from what it would be in a more ideal situation. Of course, either error still leaves us with a valid test in the sense of results being *relative* to the set of competitors. So we still have a useful test, even if the assumptions are seriously violated. It's just that we may not be able to interpret results as well as we would like.

We've been blithely tossing around "OOS performance" as if we have it in hand. Unfortunately, it's not measurable because it generally is defined in terms of an infinite population. We could approximate OOS performance by splitting our data into two parts, selecting promising competitors from one part, and estimating their OOS performance with the other part. But that's wasteful. There's a better way: cross validation.

Ordinary cross validation has a problem in many applications, including the one we are discussing. In each fold (unless we use just two folds), the in-sample (*IS*) set is much larger than the OOS set. This can skew many important families of performance statistics. Thus, we use a modified version of cross validation called *combinatorially symmetric cross validation* (CSCV).

In CSCV, we split the dataset into an even number of subsets. Then we choose half of the subsets to be the IS set, which leaves the remaining half (of equal or nearly equal number of cases) to serve as the OOS set. Repeat to cover all combinations. For example, suppose we split the data into four subsets, numbered 1, 2, 3, and 4. First we combine subsets 1 and 2 to be an IS set, leaving 3 and 4 to be the OOS set. Then we let 1 and 3 be IS, leaving 2 and 4 to be OOS. There are six such partitions possible.

For each partition, we use the IS set to find the best competitor. We also compute the OOS performance of each competitor and find the median OOS performance of all competitors. Note whether the OOS performance of the best IS performer is above the median (good news) versus less than or equal to the median (bad news). If we count the number of partitions in which the latter is true and divide this count by the total number of partitions, we get a fraction 0-1 that is an approximation to the probability that the best performer will underperform its competitors out of sample, which is a sad state of affairs. As such, we can say that this probability is a (distant) relative to the ordinary p-value that we all know and love.

Just to make this clear, suppose that the criterion we are using to judge performance is effective at capturing authentic information. In the software is available for this book, this criterion is a measure of the relationship between a single competing variable and another single variable. In the case of finding optimal parameters for a model, this criterion might be R-squared. Whatever we use, suppose for now that it is an effective measure of performance quality.

Furthermore, suppose that at least one of our competitors is truly good. In the context of this text, this means that at least one of the competing variables truly has a significant relationship with the other variable. In the context of model training (not covered here), this means that at least one of the competing finite number of parameter sets defines an effective model.

Under these two assumptions, whichever competitor has the *best* value of this criterion *in-sample* is likely *truly* the best, or at least nearly the best. Thus, we would expect its performance *out of sample* to also be exemplary. As a result, few or no partitions would find its OOS performance to be less than or equal to the median, and the computed probability would be zero or tiny.

If either of these two suppositions is violated, the situation is very different. For example, it may be that our carelessly designed criterion is a degree-ten polynomial fit that focuses heavily on noise and hence is nearly powerless at identifying truly outstanding competitors. Or it may be that all of the competitors are worthless. Maybe none of the competing variables has any relationship with the other variable. Or maybe a predictive model is fundamentally flawed, and no parameter set can make it truly work. For either type of supposition violation, IS and OOS performance will be largely unrelated and be pretty much random values. Thus, the OOS performance of the best IS performer will be all over the map, sometimes above the median and sometimes below. The IS performance has not captured anything that is indicative of OOS performance.

This discussion has focused on the best IS performer, as that is the most intuitive presentation. But it's legitimate to compute this probability for all ranks of competitors (second best, third best, etc.). If the probability is small for many of the best IS performers, then we can have considerable confidence that their performance will continue out of sample.

It may be useful to compute, for a specified even number of subsets S, how many partitions of the dataset will be involved. This is the number of combinations of S things taken $S/2$ at a time. The standard computational formula can be implemented with a simple loop, provided that the division is done in floating point rather than integer arithmetic. Here is a good way, with n_sub being the number of subsets, S, and half_S being half of that.

```
dtemp = 1.0;
for (i=0; i<half_S; i++)
    dtemp *= (double) (n_sub - i) / (double) (half_S - i);
ncombo = (int) (dtemp + 0.5);
```

The CSCV Algorithm

In this section we present the general CSCV algorithm, using C-like pseudocode. We'll use the specific application of a set of *predictor* variables competing for degree of relationship with a single other variable, called the *target* variable. However, at the appropriate points we will note how this algorithm could be easily modified for assessing the quality of parameter sets in developing a model. Also, for the sake of clarity, intuitive explanations will be liberally interspersed with the pseudocode.

First, we must be clear about how the single target variable and the set of competing predictor candidates are stored. The target is simple; it's just an array of ncases values. The predictor candidates are a bit more complicated. We have a database matrix with ncases rows and n_vars columns. However, we do not demand that all of these variables compete. We may want to ignore some of them. In fact, we will have only npreds competitors, and their column indices in the dataset are in the array preds, which is npreds long. This generalization is not needed for the algorithm, but it is convenient for the caller because it avoids the need to create a special database containing only competitors.

For convenience, here are the variables that appear often in the code:

double *dataset	Complete dataset
int ncases	Number of cases (rows) in dataset
int n_vars	Number of variables (columns) in dataset
double *all_target	All target values, ncases of them
int npreds	Number of predictors (competitors)
int *preds	Indices in database of predictors; npreds of them
int n_sub	Number of subsets, S = 2 * half_S
int half_S	Half of S
double *crits	Output
int *indices	Work vector n_sub long
int *lengths	Work vector n_sub long
int *flags	Work vector n_sub long
int *sorted_index	Work vector nvars long
double *IS_crits	Work vector nvars long
double *OOS_crits	Work vector nvars long
double *work_pred	Work vector ncases long
double *work_targ	Work vector ncases long

The first step is to partition the ncases cases in the predictor dataset and target array into n_sub (S) subsets. The array indices (n_sub long) will contain the starting index of each subset, and the corresponding array lengths will contain the number of cases in each subset. If ncases is an exact multiple of n_sub, the lengths will of course all be equal. If not, at least they should be close. Once we have these two arrays computed, it will be easy to locate the cases that correspond to each subset.

```
istart = 0;
for (i=0; i<n_sub; i++) {                    // For all S subsets
    indices[i] = istart;                     // This subset starts here
    lengths[i] = (ncases - istart) / (n_sub-i);   // It contains this many cases
    istart += lengths[i];
}
```

We have two things to initialize. Throughout the algorithm, the ncases array flags identifies whether each case is in the training set (the flag is 1) or the test set (the flag is 0). The processing of partitions begins with the first half of the subsets being the training set, and the second half the test set, so initialize accordingly. Also, the npreds array crits will count the number of times each training-set rank competitor has OOS performance less than or equal to the median. We initialize this to zero. It is a double instead of an integer because we will later convert it to a probability.

```
for (i=0; i<half_S; i++)      // This is the first partition tested
  flags[i] = 1;               // Training case
for (; i<n_sub; i++)
  flags[i] = 0;               // Test case

for (ivar=0; ivar<npreds; ivar++)
  crits[ivar] = 0.0;
```

We now begin the main outer loop that processes every partition. We don't need to know in advance how many partitions (combinations) there will be because later we'll easily know when we've done them all.

```
for (icombo=0;; icombo++) { // Main loop processes all combinations
```

The first step in this loop is to gather the in-sample targets. We count them with n. For subset ic, the cases in this subset start at indices[ic], and there are lengths[ic] of them.

```
n = 0;                                // Will count cases in the training set
for (ic=0; ic<n_sub; ic++) {          // For all S subsets of the complete dataset
  if (flags[ic]) {                    // If this subset is in the training set
    for (i=0; i<lengths[ic]; i++) {   // Get the target for this subset
      k = indices[ic]+i;              // Case index
      target[n++] = all_target[k];
    }
  }
}
```

We similarly gather the competitors in the training set. Each competitor is done individually, looping through all npreds of them. For each, ipred (supplied by the caller via preds) identifies its column in the complete dataset. Once the values for a competitor are gathered, we call compute_criterion() to compute the criterion and save the value in IS_crits. We also initialize a sort index. The call to qsortdsi() will sort the npreds criteria, simultaneously moving sorted_index so we know what's where later when we need ranks.

```
for (ivar=0; ivar<npreds; ivar++) {    // For all competitors
  n = 0;                               // Will count cases just as for target
  ipred = preds[ivar];                 // Index in complete database
  for (ic=0; ic<n_sub; ic++) {         // For all S subsets of the complete dataset
    if (flags[ic]) {                   // If this subset is in the training set
      for (i=0; i<lengths[ic]; i++) {  // Get predictor candidate for this subset
        k = indices[ic]+i;             // Case index
        competitor[n++] = dataset[k*n_vars+ipred];
      }
    }
  }

  IS_crits[ivar] = compute_criterion (n, competitor, target);
  sorted_index[ivar] = ivar;
}

qsortdsi (0, npreds-1, IS_crits, sorted_index);
```

We do exactly the same thing for the OOS cases, except that we do not sort them quite yet. First, gather the OOS targets. Then, separately for each competitor, gather those values, and compute and save the OOS criterion.

```
n = 0;                               // Will count cases in the test set
for (ic=0; ic<n_sub; ic++) {         // For all S subsets of the complete dataset
  if (! flags[ic]) {                 // If this subset is in the test set
    for (i=0; i<lengths[ic]; i++) {  // Get the target for these cases in this subset
      k = indices[ic]+i;             // Case index
      target[n++] = all_target[k];
    }
  }
}
```

```
for (ivar=0; ivar<npreds; ivar++) {      // For all competitors
   n = 0;                                 // Will count cases, just as we did above
   ipred = preds[ivar];                   // Index in complete database
   for (ic=0; ic<n_sub; ic++) {           // For all S subsets of the complete dataset
      if (! flags[ic]) {                   // If this subset is in the test set
         for (i=0; i<lengths[ic]; i++) {   // Get this competitor for this subset
            k = indices[ic]+i;             // Case index
            competitor[n++] = dataset[k*n_vars+ipred];
         }
      }
   }

   OOS_crits[ivar] = compute_criterion (n, competitor, target);
}
```

This is a good time for a brief aside on alternatives to competing for a relationship to a target variable. The basic data structure and algorithm remain the same for other alternatives. The data cases are in rows, and the competitors are in columns. For example, if the competitors are parameter sets for a model, each column represents a complete set of parameters, and each row represents the individual error for a case. In other words, the data value in row i column j would be the error for case i when parameter set j is used to define the model. Then the criterion for a collection of IS or OOS subsets would be a pooled quality measure such as R-squared for those cases.

We need to compute the median OOS performance across all competitors. There are algorithms for computing the median that are somewhat faster than sorting, but the speed of this step is inconsequential, so I take the easy way of just sorting. We must not disturb the order of the OOS criteria, so we cannot sort that array. But we no longer need the IS_crits data, because we already have the ranks via sorted_index, so we just copy the OOS criteria to the IS array and sort it to get the median.

```
for (ivar=0; ivar<npreds; ivar++)
   IS_crits[ivar] = OOS_crits[ivar];
qsortd (0, npreds-1, IS_crits);

if (npreds % 2)
   median = IS_crits[npreds/2];
else
   median = 0.5 * (IS_crits[npreds/2-1] + IS_crits[npreds/2]);
```

We just computed the median (across all competitors) of the OOS criterion. See if the OOS performance of each IS rank is less than or equal to the OOS median. Note that ivar in crits[ivar] refers to the *rank*, not the predictor index itself. For example, crits[0] refers to the *worst*-performing predictor candidate in sample in this partition, and crits[npreds-1] refers to the best IS performer, which is typically where our interest lies. Larger values of crits imply worse OOS performance.

```
for (ivar=0; ivar<npreds; ivar++) { // For all competitors
  if (OOS_crits[sorted_index[ivar]] <= median)
    ++crits[ivar];
}
```

Now we come to the real brain-buster part of the code: advancing to the next partition. Recall that we need to loop through every possible collection of $S/2$ subsets taken from the total of S subsets. Each collection will serve as the training set for a trial, with the remaining $S/2$ subsets serving as the test set. We initialized the first partition to have all $S/2$ ones first and to have the zeros last.

If you search the Internet, you will find numerous algorithms to do this, many of which are explicitly recursive. This algorithm happens to be mine, although it is possible, even likely, that someone else came up with it first and published it. Like the other algorithms that I've seen, it is recursive, but not explicitly so. I cannot offer a rigorous proof that it is correct. However, I have tested it quite thoroughly and never found it to fail.

Understanding its operation is aided by working through the code for eight partitions, writing on a sheet of paper the first dozen or two partitions. Here is the code; an intuitive explanation follows:

```
n = 0;                              // Will count 1s to we know how many to fill later
for (iradix=0; iradix<n_sub-1; iradix++) {      // Search left to right for 1-0 pattern
  if (flags[iradix] == 1) {         // Maybe; here's the 1. Count it in case we switch and fill
    ++n;                            // This many flags up to and including this one at iradix
    if (flags[iradix+1] == 0) {     // We've got the 1-0 pattern
      flags[iradix] = 0;            // Advance the 1 and replace it with a 0
      flags[iradix+1] = 1;          // Which gives us a whole new pattern
      for (i=0; i<iradix; i++) {    // Must reset everything below this change point
        if (--n > 0)                // Fill in the required number of 1s first
          flags[i] = 1;
```

```
            else                    // Then fill the rest with 0s
                flags[i] = 0;
            } // Filling in below
        break;                      // We have our new partition, so done for now
        } // If next flag is 0
      } // If this flag is 1
   } // For iradix

 if (iradix == n_sub-1) {           // True if we cannot advance to a new partition
   ++icombo;                        // Must count this last one for probability division
   break;                           // All partitions have been processed
   }
} // Main loop processes all combinations
```

The initial partition has all ones at the beginning and all zeros at the end. Each time a new partition is needed, the algorithm starts at the beginning of the flag array and searches forward, looking for the first occurrence of a one followed by a zero. The first time this pattern is encountered, the one will be shifted to the right and replaced by a zero. Not only does this give a new partition, never seen before, but any permutation of the flags prior to this pair is also unique. If this is not clear, consider that the changed pair cannot change back to one-zero and then change again to zero-one without at least one flag beyond it changing.

Once this shift has occurred, we reset all flags prior to this pair, putting the requisite number of ones at the beginning and setting the remaining flags to zero. This is where the implicit recursion enters the picture. The next time the algorithm is called upon to advance to the next partition, it will do so on a smaller subset of the flags, those to the left of the pair just switched.

Eventually the point is reached that no one-zero pairs occur inside the active area. When this happens, the rightmost one in the flag array is pushed to the right one slot, and the mass of ones has just irrevocably advanced. After the final partition (all ones on the right) appears, the one-zero pattern will no longer be found in the flag array, and we are done.

The final step is trivial: divide all criterion counts by the number of partitions to get an approximate probability that the OOS performance for each IS rank is less than or equal to the median OOS performance.

```
for (ivar=0; ivar<npreds; ivar++)
   crits[ivar] /= icombo;
```

Remember that the ivar positions in crits do not correspond to candidates but candidate *ranks*. The rankings will in general be different for different partitions. Still, it is legitimate to map these criteria to the candidates in the order of their final ranking. After we have computed the performance criteria for all candidates and ranked them, we assign the probability estimate crits[npreds-1] to whichever candidate had the best performance, and so forth, down to assigning crits[0] to the worst performer.

An Example of CSCV OOS Testing

Here is a simple example of using CSCV OOS median testing to evaluate the relationship of a set of competing candidates with a single target variable. The synthetic variables in the dataset are as follows:

- RAND0 to RAND9 are independent (within themselves and with each other) random time series.

- SUM1234 = RAND1 + RAND2 + RAND3 + RAND4

We use five-bin uncertainty reduction as our performance criterion, testing RAND0 to RAND9 as competitors to predict SUM1234. Eight CSCV subsets are used. The following results are obtained:

Variable	UncertReduc	P(<=median)
RAND4	0.0801	0.0000
RAND3	0.0784	0.0000
RAND1	0.0706	0.0000
RAND2	0.0703	0.0000
RAND5	0.0013	0.8571
RAND8	0.0012	0.8286
RAND7	0.0010	0.9000
RAND0	0.0010	0.8000
RAND6	0.0009	0.8857
RAND9	0.0006	0.7286

Not surprisingly, RAND1 to RAND4 have the highest values of uncertainty reduction. But note how extremely effective the CSCV probabilities are. The probabilities for the four variables having a true relationship are a perfect zero, while the probabilities for the unrelated variables are very high. Of course, this is a particularly easy test, but it does demonstrate the efficacy of the technique.

In my own work, I have found great value in using this CSCV algorithm to detect overfitting of the model. If you have a model that is so powerful that it is learning *noise* to the detriment of authentic patterns, you will likely find that its performance criterion is impressive, but none of the competitors has a wonderfully low CSCV probability. That's a major red flag, not to be dismissed!

Univariate Screening for Relationships

This section presents the most basic, the fastest-to-compute, and easy-to-understand technique for variable screening. In this algorithm, we have a single variable, which we call the *target*, and a (usually large) collection of variables, which we call *predictor* candidates. Usually, our application will embody this directionality, although it need not. There is nothing inherent in this algorithm that requires one variable be used to predict another. We are simply screening for a relationship.

The complete source code for this algorithm is in SCREEN_UNIVAR.CPP. It's much too long to list here in the text. At the most basic level, the algorithm is exactly as shown in the pseudocode on page 97. But there are two complications.

First, this code provides the user with a variety of relationship criteria from which to choose. Some of these require discretization into bins before processing is done, while others operate directly on continuous data. Complicating things even more is an option that is immensely valuable for extremely noisy data (such as financial market price changes). This option lets the program focus on only extreme values of the predictor candidates, those values most likely to carry predictive information, while ignoring cases that do not have extreme values. And to pile yet another complication on top of this tails-only option, every predictor candidate will have different extreme cases, so we cannot do target bin assignments based on the entire dataset. We must compute target bin thresholds separately for each candidate. This is a simple concept but very nasty coding. I won't bother discussing my code for this here; you may role your eyes at my code and choose to do it in a way that you find more comfortable. If you do want to copy my code, it's in the source file.

Another complication with this algorithm is that modern processors have multiple cores, and it would be foolish to fail to take advantage of this. My implementation is fully multithreaded, making use of every available core. Because you may be unfamiliar with methods for multithreading, I'll deal with this subject in some detail here.

One concept critical to multithreading is that a Windows thread can launch only a special function with a single parameter. Naturally, we'll need to pass a boatload of parameters to the criterion-computation routine. So, what we do is define a structure that contains all necessary parameters, fill in the contents of this structure, and then pass this structure as our solitary parameter. The structure may look something like this:

```
typedef struct {
    int varnum;      // Index of predictor (in database, not preds)
    int ncases;      // Number of cases
    int n_vars;      // Number of columns in database
    ...
    double crit;     // Criterion is returned here
} UNIVAR_CRIT_PARAMS;
```

In the calling routine, we define a variable and set as many members as possible before beginning. As threads are launched, we set any remaining parameters that could not be set until launch time, such as the ID of the variable being evaluated.

```
UNIVAR_CRIT_PARAMS univar_params[MAX_THREADS];
....
    for (ithread=0; ithread<max_threads; ithread++) {
        univar_params[ithread].ncases = n_cases;
        univar_params[ithread].n_vars = n_vars;
        ...
    }
```

On the next page, we see a C-like pseudocode outline for the entire multithreaded screening algorithm. Ideally, this will let you more easily comprehend the code in the SCREEN_UNIVAR.CPP source file. It also serves as a useful template if you want to write your own screening code from scratch.

Allocate working memory and any objects that are universally needed

Fetch all selected candidates and target from database

Perform any required initial calculations, such as finding bin boundaries and counts

For irep=0 to requested Monte-Carlo replications

Shuffle the target if we are past the first (unshuffled) replication

Allocate any objects that are dependent on the order of the targets

Set thread parameters (thread_params) that are the same for all threads

n_threads = 0   Counts the number of currently active threads
ivar = 0   Indexes (through n_candidates-1) the variable being tested
empty_slot = -1   Will be next available thread slot

Start thread loop   This is an 'endless' loop, exited only with a break

  if (ivar < n_candidates)   More variables to test?
    if (empty_slot < 0)   True while filling thread slots
      k = n_threads;
    else
      k = empty_slot;   Start this new thread in the slot recently vacated

    thread_params[k].ivar = ivar   We'll need to know which variable this is
    thread_params[k].(other stuff) = whatever   Other parms known only at launch
    threads[k] = newly created thread Launch this new thread
    ++n_threads   And count it
    ++ivar   On to the next candidate

  if (n_threads == 0)   One of two exits from the thread loop
    Break out of thread loop

The next 'if' is true if all available threads are busy and we have not yet completed launching all work

if (n_threads == max_threads && ivar < n_candidates)
  finished_id = ID of the first thread to finish OS call to wait for a thread to finish

  Next line fetches and saves the criterion for the variable just processed
  criterion[thread_params[finished_id].ivar] = thread_params[finished_id].criterion

  empty_slot = finished_id   This slot is now available
  close thread 'finished_id'
  --n_threads

Next 'if' is true if no more candidates remain to be processed
else if (ivar == n_candidates)
   Wait for all n_threads remaining threads to finish   This is a system call
   for (i=0; i<n_threads; i++)   We get here only when all threads are finished
      criterion[thread_params[i].ivar] = thread_params[i].criterion
      close thread 'i'
   Break out of thread loop   We are completely done with computation

   End of thread loop   Loop back up to top of thread loop

Free any objects that are dependent on the order of the targets

At this point, all criteria are computed and each is in crit[ivar]
Preserve and sort these for printing, and handle solo permutation test

For ivar=0 to n_candidates

  if (irep == 0)   Unpermuted runis
    sorted_crits[ivar] = original_crits[ivar] = crit[ivar]
    index[ivar] = ivar   This will let us print results sorted best to worst
    mcpt_bestof[ivar] = mcpt_solo[ivar] = 1

  else if (crit[ivar] >= original_crits[ivar])
    ++mcpt_solo[ivar]

  End of 'for all candidates' loop

For the first (unpermuted) run, sort criteria, keeping 'index' synchronized
if (irep == 0)
  Sort 'sorted_crits' ascending, simultaneously moving 'index'

else   This is a permuted run
  The next line and loop find the max criterion for this permuted run
  best_crit = criterion[0];
  For ivar=1 through n_candidates-1
    if (criterion[ivar] > best_crit)
      best_crit = criterion[ivar];
    End of 'for candidates' loop

```
    Handle the unbiased permutation test
    For ivar=0 through n_candidates-1
      if (best_crit >= original_crits[ivar])
          ++mcpt_bestof[ivar]
      End of 'for all candidates' loop

  End of MCPT replications loop
```

```
All computation is complete. Print results, sorted from max to min criterion
for (i=n_candidates-1; i>=0; i--)
  k = index[i];
  Print name, criterion, and mcpt probabilities for candidate k
  End of 'for n_candidates' counting down loop
```

```
Free all working memory and remaining objects
```

Three Simple Examples

This section demonstrates three situations, all using synthetic data to clarify the issues. The variables in the dataset are as follows:

- RAND0 to RAND9 are independent (within themselves and with each other) random time series.

- DEP_RAND0 to DEP_RAND9 are derived from RAND0 to RAND9 by introducing strong serial correlation up to a lag of nine observations. They are independent of one another.

- SUM12 = RAND1 + RAND2

- SUM34 = RAND3 + RAND4

- SUM1234 = SUM12 + SUM34

The first test run attempts to predict SUM1234 from RAND0 to RAND9, SUM12, and SUM34. The output looks like this:

114

```
--------> Mutual Information with SUM1234 <-------
```

Variable	MI	Solo pval	Unbiased pval
SUM34	0.2877	0.0001	0.0000
SUM12	0.2610	0.0001	0.0001
RAND3	0.1307	0.0001	0.0001
RAND4	0.1263	0.0001	0.0001
RAND1	0.1129	0.0001	0.0001
RAND2	0.1085	0.0001	0.0001
RAND8	0.0015	0.2994	0.9828
RAND5	0.0014	0.3673	0.9950
RAND6	0.0012	0.5303	1.0000
RAND7	0.0010	0.7384	1.0000
RAND0	0.0008	0.8332	1.0000
RAND9	0.0006	0.9605	1.0000

These results should be totally unsurprising. But do take note of the fact that the unbiased probabilities (pval) are even more indicative of the worthlessness of the worthless candidates.

The next example shows what happens when worthless and serially correlated predictors are tested with a serially correlated target. We use DEP_RAND1 to DEP_RAND9 to predict DEP_RAND0, a situation that *should* demonstrate no predictive power whatsoever. The mutual information table is as follows:

```
--------> Mutual Information with DEP_RAND0 <--------
```

Variable	MI	Solo pval	Unbiased pval
DEP_RAND2	0.0044	0.0001	0.0002
DEP_RAND4	0.0030	0.0018	0.0175
DEP_RAND3	0.0025	0.0110	0.0881
DEP_RAND6	0.0023	0.0249	0.2004
DEP_RAND9	0.0023	0.0242	0.2062
DEP_RAND8	0.0023	0.0287	0.2284
DEP_RAND1	0.0022	0.0317	0.2494
DEP_RAND5	0.0019	0.0883	0.5509
DEP_RAND7	0.0008	0.8682	1.0000

The mutual information figures are all tiny, yet the p-values show extreme significance. The careless user would surely be fooled by this, because not only are the solo p-values mostly small but even the unbiased p-value has been fooled for one or two of the candidates. This is what happens when we perform a naive statistical test on serially correlated data. Yikes.

The final example shows how the cyclic modification of the Monte Carlo permutation test can at least partially remedy the situation. We repeat the same test as that just shown, except that instead of using complete permutation, we use cyclic permutation. The results are shown here:

```
---------> Mutual Information with DEP_RAND0 <-------

     Variable        MI     Solo pval  Unbiased pval

    DEP_RAND2     0.0044      0.0513       0.3529
    DEP_RAND4     0.0030      0.2408       0.9316
    DEP_RAND3     0.0025      0.3976       0.9918
    DEP_RAND6     0.0023      0.5007       0.9976
    DEP_RAND9     0.0023      0.5237       0.9982
    DEP_RAND8     0.0023      0.4719       0.9988
    DEP_RAND1     0.0022      0.5344       0.9990
    DEP_RAND5     0.0019      0.6643       1.0000
    DEP_RAND7     0.0008      0.9920       1.0000
```

Bivariate Screening for Relationships

Sometimes a single variable acting alone has little or no predictive power, but in conjunction with another it becomes useful. The classic example is the height and weight of an individual, predicting coronary health. Either predictor alone has relatively little predictive power, but the two taken together can have great power.

Of course, in an ideal situation we could try every possible subset of predictor candidates. But this is impossible in most practical applications. In fact, for binning-type relationship criteria such as chi-square and mutual information, handling even three predictors simultaneously is often impractical because of excessively small bin counts. And the combinatoric explosion for the number of possible subsets is violent.

But two predictors at once is often a useful compromise between the simplistic weakness of just one versus the impracticality of more than two. In this section, I'll present an efficient algorithm for exhaustively screening all possible pairs of candidates. Two criteria are employed: mutual information and uncertainty reduction, although other criteria could be substituted.

We alluded to the technique used here back on page 88. Now we will be specific, showing how bin dimension unrolling can be performed efficiently. The idea is that the matrix of predictor bins is unrolled into a single vector, which itself forms one dimension of the predictor/target bin matrix. For example, suppose the two predictors are each split into three bins, and the target is split into four. The unrolled predictor dimension would consist of 3×3=9 bins, meaning that we perform the analysis with a 9 by 4 matrix.

The algorithm presented has an interesting bonus feature: it allows the user to specify multiple target candidates. The algorithm will optionally find individual targets that have maximum predictability from associated bivariate pairs of predictors. One example of the utility of multiple target candidates is when the application is predicting future movement of a financial market with the goal of taking a position and then ideally closing the position with a profit. Should we employ a tight stop to discourage severe losses? Or should we use a loose stop to avoid being closed out by random noise? We might test multiple targets corresponding to various degrees of stop positioning and then determine which of the competitors is most predictable.

The easiest way to present the complete algorithm is to break it into sections, sometimes showing exact code and sometimes just an outline. We begin with an outline of the overall process, with special emphasis on the Monte Carlo permutation tests. You might want to review that prior section, especially the material on selection bias that begins on page 95.

Compute *n_combo* as the total number of combinations of predictors and target candidates.

Allocate working memory and any objects that are universally needed

Fetch all selected predictor and target candidates from database

Perform any required initial calculations, such as finding bin boundaries, counts, and marginals

```
for (irep=0; irep<mcpt_reps; irep++) {

   Shuffle target if in permutation run (irep>0)

   Compute and save criterion for all combinations (done with bivar_threaded())

   for (icombo=0; icombo<n_combo; icombo++) { // Update the MCPT

      if (icombo == 0 || crit[icombo] > best_crit)
         best_crit = crit[icombo];

      if (irep == 0) {       // Original, unpermuted data
         original_crits[icombo] = crit[icombo];
         mcpt_bestof[icombo] = mcpt_solo[icombo] = 1;
      }

      else if (crit[icombo] >= original_crits[icombo])
         ++mcpt_solo[icombo];

      } // For all combinations

   if (irep > 0) {
      for (icombo=0; icombo<n_combo; icombo++) {
         if (best_crit >= original_crits[icombo]) // Valid only for largest
            ++mcpt_bestof[icombo];
      }
      } // If irep>0

   } // For all MCPT replications

All computation is finished. Print.
Clean up and exit.
```

The algorithm shown here is similar to that presented on page 88. The nitty-gritty computation is done in subroutine bivar_threaded(), which we'll soon explore. The complete source code can be found in the file SCREEN_BIVAR.CPP. But let's begin with the routine for computing mutual information. This is a bin-unrolled version of the most basic definition of mutual information, shown in Equation (1.16) on page 18.

```
static double compute_mi (
   int ncases,              // Number of cases
   int nbins_pred,          // Number of predictor bins
   int *pred1_bin,          // Ncases vector of predictor 1 bin indices
   int *pred2_bin,          // Ncases vector of predictor 2 bin indices
   int nbins_target,        // Number of target bins
   int *target_bin,         // Ncases vector of target bin indices double
   *target_marginal,        // Target marginal
   int *bin_counts          // Work area nbins_pred_squared*nbins_target long
   )
{
   int i, j, k, nbins_pred_squared;
   double px, py, pxy, MI;

   // Zero all bin counts

   nbins_pred_squared = nbins_pred * nbins_pred; // Predictor bins unrolled

   for (i=0; i<nbins_pred_squared; i++) {
      for (j=0; j<nbins_target; j++)
         bin_counts[i*nbins_target+j] = 0;
      }

   // Compute bin counts for bivariate predictor and full table

   for (i=0; i<ncases; i++) {
      k = pred1_bin[i]*nbins_pred+pred2_bin[i];     // Index in unrolled predictor array
      ++bin_counts[k*nbins_target+target_bin[i]];   // Bin in predictor/target matrix
      }

   // Compute mutual information

   MI = 0.0;
   for (i=0; i<nbins_pred_squared; i++) {          // Unrolled predictor bins
      k = 0;
      for (j=0; j<nbins_target; j++) // Sum across target bins to get predictor marginal
         k += bin_counts[i*nbins_target+j];
      px = (double) k / (double) ncases;
```

```
    for (j=0; j<nbins_target; j++) {
      py = target_marginal[j];
      pxy = (double) bin_counts[i*nbins_target+j] / (double) ncases;
      if (pxy > 0.0)
        MI += pxy * log (pxy / (px * py));        // Equation (1.16) on Page 18
      }
    }

  if (nbins_pred_squared <= nbins_target)
    MI /= log ((double) nbins_pred_squared); // Normalize 0-1
  else
    MI /= log ((double) nbins_target);

  return MI;
}
```

This code assumes that both predictors are split into the same number of bins. This restriction is not necessary in general; it's just a programming convenience for this demonstration. Thus, the number of unrolled predictor bins is the number of individual bins squared. Also, for easier user interpretability, the mutual information is divided by its maximum possible value, which normalizes the quantity to the range 0-1.

Last, we'll explore the core of this algorithm, the subroutine that computes the criteria for all possible pairs of predictors and individual target candidates. As we've seen in prior multithreading examples, we need a data structure through which all parameters are passed to the threaded routine. It's straightforward, so we'll dispense with listing it or the trivial wrapper routine here; see SCREEN_BIVAR.CPP for a complete listing. Instead, we focus only on bivar_threaded(). Shown next is the basic listing, with error handling and other extraneous code omitted for clarity.

Pay attention to the fact that when we initialize the parameter-passing structure, each thread gets its own private bin_counts and bivar_counts work areas.

The trickiest part of this code is the short section with the comment Advance to the next combination on page 122. This counts up through all possible trios of two predictors and one target, with the target changing fastest. Study it.

```
static int bivar_threaded (
  int max_threads,          // Maximum number of threads to use
  int ncases,               // Number of cases
  int npred,                // Number of predictor candidates
```

```
   int ntarget,                 // Number of target candidates
   int nbins_pred,              // Number of predictor bins
   int *pred_bin,               // Ncases vector of predictor bin indices, npred of them
   int nbins_target,            // Number of target bins
   int *target_bin,             // Ncases vector of target bin indices, ntarget of them
   double *target_marginal,     // Target marginal, ntarget of them
   int which,                   // 1=mutual information, 2=uncertainty reduction
   double *crit,                // Output of all criteria, npred*(npred-1)/2*ntarget long
   int *bin_counts,             // Work area
                                // max_threads*nbins_pred*nbins_pred*nbins_target
   int *bivar_counts            // Work area max_threads*nbins_pred_squared long
   )
{
   int i, k, ret_val, ithread, n_threads, empty_slot;
   int ipred1, ipred2, itarget, icombo, n_combo;
   BIVAR_PARAMS bivar_params[MAX_THREADS];
   HANDLE threads[MAX_THREADS];

/*
   Initialize those thread parameters which are constant for all threads.
   Each thread will have its own private bin_count and bivar_count matrices for working storage.
   They must not share scratch storage!
*/

   for (ithread=0; ithread<max_threads; ithread++) {
     bivar_params[ithread].ncases = n_cases;
     bivar_params[ithread].nbins_pred = nbins_pred;
     bivar_params[ithread].nbins_target = nbins_target;
     bivar_params[ithread].bin_counts = bin_counts +
                                 ithread * nbins _pred * nbins_pred * nbins_target;
     bivar_params[ithread].bivar_counts = bivar_counts +
                                 ithread * nbins _pred * nbins_pred;
     bivar_params[ithread].which = which;
     } // For all threads, initializing constant stuff
```

```
/*
   Do it
   We use icombo to define a unique set of two predictors and one target.
   It ranges from 0 through npred * (npred-1) / 2 * ntarget.
*/

   n_threads = 0;                     // Counts threads that are active
   for (i=0; i<max_threads; i++)
      threads[i] = NULL;              // Thread pointers

   // The first trio is the first predictor candidate, the second, and the first target

   ipred1 = itarget = icombo = 0     // icombo will encode the trio being processed
   ipred2 = 1;

   n_combo = npred * (npred-1) / 2 * ntarget; // This many combinations

   empty_slot = -1; // After full, will identify the thread that just completed
   for (;;) {         // Main thread loop processes all predictors

/*
   Start a new thread if we still have work to do
*/

      if (icombo < n_combo) {      // If there are still some trios to do
         if (empty_slot < 0)       // Negative while we are initially filling the queue
            k = n_threads;         // This is the next available slot
         else                      // The queue has been filled and running
            k = empty_slot;        // The most recently completed slot, now available

      bivar_params[k].icombo = icombo;  // Needed for placing final result
      bivar_params[k].pred1_bin = pred_bin+ipred1*ncases;
      bivar_params[k].pred2_bin = pred_bin+ipred2*ncases;
      bivar_params[k].target_bin = target_bin+itarget*ncases;
      bivar_params[k].target_marginal = target_marginal+itarget*nbins_target;
      threads[k] = (HANDLE) _beginthreadex (NULL, 0, bivar_threaded_wrapper,
                                        &biv ar_params[k], 0, NULL);

      ++n_threads;
```

```
// Advance to the next combination; itarget changes fastest, ipred1 slowest

++icombo;
if (itarget < ntarget-1)
  ++itarget;
else {
  itarget = 0;
  if (ipred2 < npred-1)
    ++ipred2;
  else {
    ++ipred1;
    ipred2 = ipred1 + 1;
    }
  }

} // if (icombo < n_combo), meaning that we have more work to do

if (n_threads == 0) // Are we done?
  break;
/*
Handle full suite of threads running and more threads to add as soon as some are done.
Wait for just one thread to finish. Feel free to change the 500000 timeout.
*/

if (n_threads == max_threads && icombo < n_combo) {
  ret_val = WaitForMultipleObjects (n_threads, threads, FALSE, 500000);
  crit[bivar_params[ret_val].icombo] = bivar_params[ret_val].crit;
  empty_slot = ret_val;     // Index of thread that just finished
  CloseHandle (threads[empty_slot]);
  threads[empty_slot] = NULL;
  --n_threads;
  }
```

```
/*
   Handle all work has been started and now we are just waiting for threads to finish
*/

    else if (icombo == n_combo) {
      ret_val = WaitForMultipleObjects (n_threads, threads, TRUE, 500000);
      for (i=0; i<n_threads; i++) {
        crit[bivar_params[i].icombo] = bivar_params[i].crit;
        CloseHandle (threads[i]);
        }
      break;
      }
    } // Endless loop which threads computation of criterion for all predictors

  return 0;
}
```

In the routine just listed, work can be roughly divided into three blocks. The first block (if (icombo < n_combo)) checks to see whether there is still work to do. If so, it launches a new thread. The second block (if (n_threads == max_threads && icombo < n_combo)) is executed if all threads are busy and there is still work to do. It just sits and waits for a thread to finish. The third block (else if (icombo == n_combo)) is executed just once, when all work has been launched. It sits and waits for *all* threads to finish.

Stepwise Predictor Selection Using Mutual Information

In the prior chapter, you learned what mutual information is, why it is important, and how to compute it. In the prior section you saw how it (and other criteria) can be used to screen for *individual* relationships between a collection of candidates and a single target variable. Now you will learn how to use it intelligently to select a predictor variable *set* that is likely to be effective. This can be enormously valuable when you have a massive number of candidates and need to whittle this universe down to a manageable number before embarking on expensive training of sophisticated models. In particular, we will explore two specific algorithms that employ highly effective stepwise predictor selection.

Maximizing Relevance While Minimizing Redundancy

Let X_1, X_2, ..., X_M be a set of predictor candidates for predicting Y. Given some $m<M$, we want to find m members of this collection such that this subset, which we call S, has maximum *joint dependency* with Y. Joint dependency is an extension of mutual information in which one of the quantities is a collection of random variables rather than a single random variable. We can think of the joint dependency as the mutual information between S and Y, $I(S;Y)$. For convenience, let S be the first m candidates. Then this joint dependency is given by Equation (2.13), a straightforward extension of Equation (1.23).

$$I(S;Y) = \int \cdots \int \; f_{S,Y}(x_1,...,x_m,y)$$
$$\log \frac{f_{S,Y}(x_1,...,x_m,y)}{f_S(x_1,...,x_m)f_Y(y)} dx_1,...,dx_m dy \tag{2.13}$$

Unfortunately, in practice this quantity is impossible to compute for $m>2$ and is often difficult even for $m=2$. The reason is that the multiple integration involves implicitly or explicitly partitioning the dataset in more than two dimensions, leading to excessive thinning of the density approximations. Consider the simplest case of $m=2$. Suppose there are 1,000 cases. We have a rectangular checkerboard for the two predictors, and we have a stack of these checkerboards to accommodate Y. Each case will have a position in this three-dimensional cube. If we were to partition each dimension into ten bins, we would have $10^3=1000$ bins, leading to an average of just one case per bin. If $m=3$, there would be an average of one-tenth of a case per bin! Clearly, there is no hope of implementing the direct approach to finding the optimal subset S if $m>2$, and there's probably no hope even for $m=2$ unless there are an enormous number of cases. The density approximations that are critical to the integrand are simply too inaccurate.

There is another problem, too. Combinatoric explosion is a standard nemesis of any predictor selection algorithm. If we are choosing m of M candidates, there are $M!/(m!(M-m)!)$ possible combinations. This is often so large that trying all of them is out of the question. A shortcut is needed.

There are several shortcuts in use, the most important of which were discussed earlier in this chapter. To briefly review, the simplest and most common is *first-order incremental search*, more commonly called *forward stepwise selection*. We first choose the single best predictor, where "best" is defined in terms of some ideally intelligent criterion. Then we find the predictor that, when combined with the first, produces the maximum increment in whatever performance criterion is being evaluated. A third is added in the same way, and so forth.

It is theoretically possible for this method to fail, perhaps miserably. Suppose, for example, that variables 21 and 35 together do a superb job of predicting Y, although neither alone is any good. Maybe variable 17 is the best single predictor, while variable 19 provides the best incremental power. These two variables together may not come even close to being as good as 21 and 35. This is sad but often unavoidable.

Other techniques do exist. Higher-order methods keep not just the best variable at each step but several of the best, which increases the likelihood of finding the optimal set. Backward selection starts by using all candidates and removing one at a time. However, first-order incremental search is the most efficient, making it the only practical choice in any application in which computational resources are limited. This is the approach used here, not only because of its efficiency but because of a fortuitous property of the algorithm when applied to joint dependency.

Peng, Long, and Ding (2005), in their paper "Feature Selection Based on Mutual Information: Criteria of Max-Dependency, Max-Relevance, and Min Redundancy," provide a selection algorithm that is simple, elegant, and almost miraculously duplicates first-order incremental optimization of Equation (2.13), without ever having to evaluate the equation. I now present an intuitive development of this algorithm.

The *relevance* of a set of predictors S to a predicted variable Y is defined as the mean mutual information between Y and each predictor in S. This is shown in Equation (2.14), where |S| is the number of predictors in the set.

$$Relevance(Y,S) = \frac{1}{|S|} \sum_{X_i \in S} I(Y;X_i) \tag{2.14}$$

It is tempting to simply maximize this quantity. We would begin by selecting the single predictor that has maximum mutual information with Y. Then we add the candidate that has second-max mutual information, and so forth, until we have m predictors in S. This would obviously maximize the relevance of S.

The problem with this simplistic approach is that it ignores the fact that S chosen this way will usually contain an enormous amount of redundancy. If two variables have high mutual information with Y, chances are they also have high mutual information with each other. It will probably be the case that if we simply choose a new variable that has high mutual information with Y, appending it to S will not improve the joint dependency between S and Y very much because it won't be add much information that is new.

The algorithm of [Peng, Long, and Ding, 2005] solves this problem by choosing the next variable as the one having maximum value of its mutual information with Y, minus its redundancy with the existing set of predictors. The definition of redundancy is shown in Equation (2.15). Note that the redundancy of a predictor candidate with S is the same as the relevance of this candidate with S. The only difference is the name of the quantity. The term *relevance* is used when referring to the predicted variable, while *redundancy* is used when referring to another predictor candidate.

$$Redundancy\left(X_j;S\right) = \frac{1}{|S|}\sum_{X_i \in S} I\left(X_j;X_i\right) \tag{2.15}$$

In summary, the algorithm begins by choosing the single predictor that has maximum mutual information with Y. Let S be this one variable. From then on, we add one new variable at a time by choosing the one that maximizes the criterion shown in Equation (2.16), stopping when we have the desired number m of predictors in S.

$$Criterion\left(X_j;S\right) = I\left(X_j;Y\right) - \frac{1}{|S|}\sum_{X_i \in S} I\left(X_j;X_i\right) \tag{2.16}$$

This algorithm makes obvious intuitive sense. At each step we want to simultaneously maximize the mutual information with Y while minimizing the average mutual information with the predictors already in S. What is not at all obvious is that this algorithm will choose exactly the same variables as would be chosen if we were able to evaluate Equation (2.13), something that we have already seen to be practically impossible. The proof can be found in the original paper. All we do here is marvel that we can capitalize on this extraordinary result.

There are two Monte Carlo permutation tests that can be performed as this algorithm executes. We can do a "solo" test by comparing the relevance of each individual candidate to its permuted values. This provides straightforward individual candidate significance tests. We can also, as each new variable is added to the "kept" set, test the significance of the "so-far" collection of variables. This is done by cumulating the sum of the individual relevances and comparing this sum to the corresponding values under permutation. For each quantity of kept variables, this provides the estimated probability that if the variables were *all* worthless, we could have achieved this much total relevance by sheer good luck.

Code for the Relevance Minus Redundancy Algorithm

The file SCREEN_RR.CPP contains a subroutine that implements the Peng-Long-Ding algorithm for relevance-minus-redundancy predictor selection. Rather than list it all in its complex glory, I'll just provide a C-like outline of the algorithm stripped down to the bare essentials. This should be sufficient for you to produce your own custom implementation. The complete source file will fill in additional details, if needed. Here it is, with comments interspersed:

```
Allocate working memory and any objects that are universally needed

Fetch all selected candidates and target from database

Perform any required initial calculations, such as finding bin boundaries and marginals
```

This is the main outermost loop for the Monte Carlo permutation test:

```
for (irep=0; irep<mcpt_reps; irep++) {

  Shuffle target if in permutation run (irep>0)
```

Here we call a subroutine that uses multithreading to compute the mutual information between each individual candidate and the target.

```
  First step: Compute and save (in crit) MI criterion for all individual candidates
```

We save this set of mutual information measures in relevance because they will be needed later, as we add new predictors to the kept set.

This will be the first term in Equation (2.16). Also, we find the maximum mutual information criterion among competitors.

```
  for (ivar=0; ivar<npred; ivar++) {
    relevance[ivar] = crit[ivar];  // Will need this for Step 2, addition of more predictors
    if (ivar == 0 || crit[ivar] > best_crit) {
      best_crit = crit[ivar];
      best_ivar = ivar;
      }
    }
```

We keep in stepwise_crit and stepwise_ivar a record of the variables and associated criterion as they are added. We just found the first, so its subscript is zero. Also, sum_relevance will cumulate the total relevance of the kept set. This plays no role whatsoever in the selection algorithm. Its sole purpose is to permit a Monte Carlo permutation test of the "so-far" significance of the kept set.

```
stepwise_crit[0] = best_crit;      // Criterion for first var is largest MI
stepwise_ivar[0] = best_ivar;      // It's this candidate
sum_relevance = best_crit;         // Will cumulate as more vars added
```

If this is the first (unpermuted) replication, then we preserve the "original" values of these quantities. We also initialize the count for the so-far permutation test. Then we preserve the original relevance and criterion (which are equal for step 1, the first variable) and initialize the counts for each solo permutation test. Finally, this would be a good place to print for the user a table of these first-step criteria, the mutual information of each candidate with the target.

```
if (irep == 0) {   // Original, unpermuted data

  original_stepwise_crit[0] = best_crit;   // Criterion for first var is largest MI
  original_stepwise_ivar[0] = best_ivar; // It's this candidate
  original_sum_relevance[0] = sum_relevance;
  stepwise_mcpt_count[0] = 1;         // Initialize cumulative MCPT

  for (ivar=0; ivar<npred; ivar++) {
    original_relevance[ivar] = current_crits[ivar] = crit[ivar];
    solo_mcpt_count[ivar] = 1;        // Initialize solo MCPT
    }

  Print sorted table of individual MIs
  } // If irep=0 (original, unpermuted run)
```

If we are no longer in the unpermuted replication, then we have to handle the two permutation tests. The "stepwise" test is for the collection of variables so far, which of course is just one, the single best, at this time. The "solo" test is done separately for each candidate, individually.

```
else {                              // Count for MCPT
  if (sum_relevance >= original_sum_relevance[0])
    ++stepwise_mcpt_count[0];
  for (ivar=0; ivar<npred; ivar++) {
    if (relevance[ivar] >= original_relevance[ivar])
      ++solo_mcpt_count[ivar];
  }
} // Permuted replication
```

At this time, we have computed and saved in relevance the mutual information of each candidate with the target, and we have selected the best for inclusion in the "kept" set. Now we iteratively add more candidates. Note that the redundancy of a candidate can change as predictors are added. This is because the kept set is increasing, so their mean redundancy changes. We will keep in sum_redundancy[] the *total* redundancy of each remaining candidate with the variables in the "kept" set. Initialize this to zero for all npred candidates.

```
for (i=0; i<npred; i++)
  sum_redundancy[i] = 0.0;

for (nkept=1; nkept<max_pred; nkept++) { // Main 'adding' loop

  Print candidates kept so far (if in unpermuted rep)
```

Build in which_preds the k candidates not yet selected. This code is not shown here because although it is simple, it is distracting. See SCREEN_RR.CPP for the details of how I do it. Then call a routine (rr_threaded()) that uses multithreading to compute the mutual information between the variable just added and each of the remaining candidates (which_preds). These are placed in crit[] so we can soon update the redundancies.

A long time ago, we saved in relevance the first term in Equation (2.16). A moment ago we computed one member of the summation in the right term of this equation. We now update that sum and evaluate Equation (2.16) to get the criterion for each remaining candidate variable. Find the candidate with the maximum criterion.

```
for (i=0; i<npred-nkept; i++) { // Cumulate sum redundancy, then compute criteria
   k = which_preds[i];  // Index in preds of this candidate
   sum_redundancy[k] += crit[i];
   current_crits[i] = relevance[k] - sum_redundancy[k] / nkept; // Equation (2.16)
   if (i == 0 || current_crits[i] > best_crit) {
      best_crit = current_crits[i];
      best_ivar = k;
      }
   }
```

Preserve the best candidate and its criterion. Also sum the relevance for the "so-far" permutation test.

```
stepwise_crit[nkept] = best_crit;
stepwise_ivar[nkept] = best_ivar;
sum_relevance += relevance[best_ivar];
```

If we are in the unpermuted replication, save these quantities for later printing and comparisons on which the permutation tests are based. Otherwise, do the counting for the permutation test.

```
if (irep == 0) {        // Original, unpermuted
   original_stepwise_crit[nkept] = best_crit;
   original_stepwise_ivar[nkept] = best_ivar;
   original_sum_relevance[nkept] = sum_relevance;
   stepwise_mcpt_count[nkept] = 1;
   }
else {                  // Count for MCPT
   if (sum_relevance >= original_sum_relevance[nkept])
      ++stepwise_mcpt_count[nkept];
   } // Permuted
   } // Second step (for nkept): Iterate to add predictors to kept set
} // For all MCPT replications
```

That's it. We can now print a table of final results and then free any objects and memory that were allocated at the start of this routine.

An Example of Relevance Minus Redundancy

This section demonstrates a revealing example of the algorithm using synthetic data to clarify the presentation. The variables in the dataset are as follows:

- RAND0 to RAND9 are independent (within themselves and with each other) random time series.

- SUM12 = RAND1 + RAND2

- SUM34 = RAND3 + RAND4

- SUM1234 = SUM12 + SUM34

The test run attempts to predict SUM1234 from RAND0 to RAND9, SUM12, and SUM34. The output is shown here, with comments interspersed:

```
***************************************************************
*                                                             *
*    Relevance minus redundancy for optimal predictor subset  *
*      12 predictor candidates                                *
*      12 best predictors will be printed                     *
*       5 predictor bins                                      *
*       5 target bins                                         *
*     100 replications of Monte-Carlo Permutation Test        *
*                                                             *
***************************************************************

Initial candidates, in order of decreasing mutual information with SUM1234

          Variable          MI
            SUM34         0.2877
            SUM12         0.2610
            RAND3         0.1307
            RAND4         0.1263
            RAND1         0.1129
            RAND2         0.1085
            RAND8         0.0015
            RAND5         0.0014
            RAND6         0.0012
```

RAND7	0.0010		
RAND0	0.0008		
RAND9	0.0006		

Predictors so far	Relevance	Redundancy	Criterion
SUM34	0.2877	0.0000	0.2877

We see from the previous table that the first candidate chosen is the one that has maximum mutual information with the target. Naturally this would be either SUM12 or SUM34, and it happens to be the latter. Then, in the following table we see that SUM12 has the largest relevance (its mutual information with the target) and essentially no redundancy with SUM34 (again, no surprise). This gives it the highest selection criterion, and it is chosen.

Additional candidates, in order of decreasing relevance minus redundancy

Variable	Relevance	Redundancy	Criterion
SUM12	0.2610	0.0014	0.2596
RAND1	0.1129	0.0016	0.1112
RAND2	0.1085	0.0009	0.1076
RAND6	0.0012	0.0007	0.0005
RAND0	0.0008	0.0009	-0.0000
RAND8	0.0015	0.0017	-0.0002
RAND5	0.0014	0.0016	-0.0002
RAND9	0.0006	0.0008	-0.0002
RAND7	0.0010	0.0012	-0.0003
RAND3	0.1307	0.3154	-0.1847
RAND4	0.1263	0.3158	-0.1895

Predictors so far	Relevance	Redundancy	Criterion
SUM34	0.2877	0.0000	0.2877
SUM12	0.2610	0.0014	0.2596

Now we come to an important observation. One might think that the next candidate selected would be either RAND1, RAND2, RAND3, or RAND4, which are the four components of the SUM1234 target. However, the table on the next page shows that these four candidates actually fall at the bottom of the list! This is because they have so much redundancy with SUM12 and SUM34 (taken as a group) that they will not be chosen next. In fact, RAND6, which has no relationship whatsoever with any of the other variables, is chosen based only on its tiny random relevance and slightly smaller random redundancy.

Additional candidates, in order of decreasing relevance minus redundancy

Variable	Relevance	Redundancy	Criterion
RAND6	0.0012	0.0009	0.0003
RAND0	0.0008	0.0008	0.0000
RAND8	0.0015	0.0015	0.0000
RAND9	0.0006	0.0008	-0.0002
RAND5	0.0014	0.0017	-0.0003
RAND7	0.0010	0.0013	-0.0004
RAND3	0.1307	0.1581	-0.0274
RAND4	0.1263	0.1585	-0.0322
RAND1	0.1129	0.1527	-0.0398
RAND2	0.1085	0.1485	-0.0399

Predictors so far	Relevance	Redundancy	Criterion
SUM34	0.2877	0.0000	0.2877
SUM12	0.2610	0.0014	0.2596
RAND6	0.0012	0.0009	0.0003

But now that the selected set's redundancy with the remaining candidates has been "diluted" by the inclusion of the unrelated RAND6, RAND1 to RAND4 jump to the top of the list because of their relatively large relevance but lessened redundancy.

Additional candidates, in order of decreasing relevance minus redundancy

Variable	Relevance	Redundancy	Criterion
RAND3	0.1307	0.1058	0.0249
RAND4	0.1263	0.1061	0.0202
RAND1	0.1129	0.1021	0.0107
RAND2	0.1085	0.0995	0.0090
RAND0	0.0008	0.0010	-0.0002
RAND9	0.0006	0.0009	-0.0003

RAND5	0.0014	0.0017	-0.0003
RAND8	0.0015	0.0018	-0.0004
RAND7	0.0010	0.0015	-0.0006

Predictors so far	Relevance	Redundancy	Criterion
SUM34	0.2877	0.0000	0.2877
SUM12	0.2610	0.0014	0.2596
RAND6	0.0012	0.0009	0.0003
RAND3	0.1307	0.1058	0.0249

There is little point in continuing to show the inclusion steps. We now jump to the final table that lists all candidates in the order in which they were selected, along with associated p-values.

----------> Final results predicting SUM1234 <----------

Preds	Relevance	Redundancy	Criterion	Solo pval	Group pval
SUM34	0.2877	0.0000	0.2877	0.010	0.010
SUM12	0.2610	0.0014	0.2596	0.010	0.010
RAND6	0.0012	0.0009	0.0003	0.570	0.010
RAND3	0.1307	0.1058	0.0249	0.010	0.010
RAND4	0.1263	0.0797	0.0465	0.010	0.010
RAND1	0.1129	0.0617	0.0511	0.010	0.010
RAND2	0.1085	0.0505	0.0581	0.010	0.010
RAND8	0.0015	0.0014	0.0001	0.320	0.010
RAND5	0.0014	0.0014	-0.0001	0.340	0.010
RAND7	0.0010	0.0014	-0.0004	0.650	0.010
RAND0	0.0008	0.0013	-0.0004	0.850	0.010
RAND9	0.0006	0.0012	-0.0006	0.980	0.010

Two different p-values are printed for each predictor candidate. The *Solo pval* is the same quantity printed in the univariate test (page 110). This is the probability that if this predictor has no actual mutual information with the target, a mutual information (relevance here) as large as that obtained could have occurred. Understand that this quantity considers each candidate in isolation, not involving any other candidates. Note how nicely this reveals the uselessness of the third candidate chosen, RAND6.

The *Group pval* considers the associated candidate along with *every prior candidate*. It tests the null hypothesis that the *group* of candidates selected so far, on average, has no mutual information with the target.

Regrettably, I am not aware of any way of computing what would be an especially useful p-value—one that tests the null hypothesis that selecting the candidate provides no additional (nonredundant) relevance. Such a p-value would be valuable for determining when to stop including additional candidates in the selected subset. The problem appears to be that the test statistic at any step is strongly dependent on the relevance of those predictors already selected. If anyone knows of a way around this problem, I would love to hear about it.

A Superior Selection Algorithm for Binary Variables

If the predicted variable and all predictor candidates are binary, then we can use a stepwise selection algorithm that seems to be superior to the *PLD* algorithm (presented by F. Fleuret in the 2004 paper "Fast Binary Feature Selection with Conditional Mutual Information"). Recall that the *PLD* algorithm has the fabulous property that its selections are identical to those that would be obtained by forward stepwise selection based on the optimal but impossible Equation (2.13). Nonetheless, also recall that forward stepwise selection is itself suboptimal. The optimal method is to examine every possible combination of predictors, a task that is usually impractical, even if we could evaluate the criterion of Equation (2.13), which of course we cannot. So, there is room for improvement.

Actually, the Fleuret algorithm described in this section can theoretically be used for any discrete variables, not just binary. It's just that unless the number of cases is huge, the algorithm fails because of sparse bins. For this reason, it is typically implemented only for binary data.

We need to introduce the notion of conditional mutual information. Recall from Equation (1.13) on page 18 that the mutual information shared by two variables is equal to the entropy of one of them minus its entropy conditional on the other. This is shown in Equation (2.17). Intuitively, this means that the information shared by X and Y is equal to the information in Y minus the information content of Y that is above and beyond that provided by X. Equivalently, the total information in Y is equal to that which is shared with X plus that which is above and beyond X.

$$I(X;Y) = I(Y;X) = H(Y) - H(Y|X) \tag{2.17}$$

Now suppose that we already possess some information in the form of the value of some variable Z. We can then talk about the mutual information of X and Y given that we know Z, written as $I(X;Y|Z)$. If Z happens to be totally unrelated to X and Y, its knowledge will have no impact on the mutual information of X and Y. At the other extreme, it may be that X and Y share a lot of information, but Z happens to completely duplicate this shared information. In this case, $I(X;Y)$ will be large, but $I(X;Y|Z)$ will be zero. Conditional mutual information can be computed with Equation (2.18). Observe that this is a simple extension of Equation (2.17), obtained by conditioning all terms on Z.

$$I(X;Y|Z) = I(Y;X|Z) = H(Y|Z) - H(Y|X,Z)$$ (2.18)

Conditional mutual information allows us to approach the problem of redundancy from a different direction. Recall from the *PLD* algorithm that our goal is to find a variable from among the candidates that has high mutual information with Y and low joint mutual information with the predictors already selected. We now have an excellent tool. Suppose X is a candidate for inclusion and Z is a variable that is already in S, the set of predictors chosen so far. The conditional mutual information of X and Y given Z measures how much the candidate X contributes to predicting Y above and beyond what we already get from Z. A good candidate will have a large value of $I(X;Y|Z)$ for every Z in S. If there is even one variable Z in S for which $I(X;Y|Z)$ is small, there is little point in including this candidate X, because it contributes little beyond what is already contributed by that Z. This inspires us to choose the candidate X that has the maximum value of the criterion shown in Equation (2.19).

$$Criterion(X,Y,S) = \min_{Z \in S} I(X;Y|Z)$$ (2.19)

Equation (2.18) is a good intuitive definition of conditional mutual information, but it is not the easiest way to compute it. A better way is Equation (2.20).

$$I(X;Y|Z) = H(X,Z) + H(Y,Z) - H(Z) - H(X,Y,Z)$$ (2.20)

The file MUTINF_B.CPP contains the complete source code to evaluate this equation for X, Y, and Z arrays. This code is simple but very tedious, so I will not reproduce it in its entirety here. The easiest approach, though not necessarily the most efficient, is to use nested logical expressions to tally the two-by-two-by-two bin counts. This is done as shown here:

```
n000 = n001 = n010 = n011 = n100 = n101 = n110 = n111 = 0;
for (i=0; i<n; i++) {
  if (x[i]) {
    if (y[i]) {
      if (z[i])
        ++n111;
      else
        ++n110;
    }
    else {
      if (z[i])
        ++n101;
      else
        ++n100;
    }
  }
  else {
    if (y[i]) {
      if (z[i])
        ++n011;
      else
        ++n010;
    }
    else {
      if (z[i])
        ++n001;
      else
        ++n000;
    }
  }
}
```

Once the eight bins counts are tallied, computing the four terms in Equation (2.20) is straightforward. For example, $H(Z)$ can be computed with the following code:

```
nz0 = n000 + n010 + n100 + n110;
nz1 = n - nz0;
if (nz0) {
  p = (double) nz0 / (double) n;
  HZ = p * log (p);
  }
else
  HZ = 0.0;
if (nz1) {
  p = (double) nz1 / (double) n;
  HZ += p * log (p);
  }
```

The other four terms are computed similarly. See the code for details. It should be noted that [Fleuret, 2004] discusses faster ways of summing the bin counts. Since the variables are all binary, values of X, Y, and Z can be encoded as bits in integers. By using logical conjunctions of these integers, along with table lookups, the bin counts can be found very quickly. I have not found speed to be a problem, so I have not implemented this algorithm.

The interesting part of the variable selection procedure is the stepwise algorithm. We begin by selecting the candidate that has maximum mutual information with Y. After that, for each step we evaluate the criterion of Equation (2.19) for each remaining candidate and choose the candidate having the greatest criterion. However, there is more to consider...

Fleuret describes a cute trick for avoiding having to check every candidate against every Z, which can consume enormous amounts of time if there are a lot of variables in the kept set S. When a new Z is tested in computing the minimum across all Zs in S, the minimum obviously cannot increase. So if the minimum across Z so far is already less than the best candidate criterion so far, there is no point in continuing to test more Zs for the candidate. This candidate has already lost the competition for this round. Of course, we need to keep track of, for each candidate, the place where we have stopped testing it against Zs. This is because on a later round of adding a variable, the best so far may be small, and a candidate whose testing was stopped early on a prior round may need to be tested against more Zs to see whether it might be the best now. *A tentative winner cannot be confirmed until it has been checked for all Zs, but a loser can be eliminated early.*

Stepwise selection of predictor variables using the Fleuret algorithm is quite similar to routines already presented, so we will not examine it in detail here. Also, a complete implementation is available in the file MI_BIN.CPP. However, examination of a simplified snippet helps to understand proper implementation of the algorithm.

The loop shown in the following code is invoked after one variable, that having maximum mutual information with *Y*, has been picked. At this time, scores[icand] has been initialized to the mutual information between that candidate and *Y*, and last_indices[icand] has been initialized to –1 for all candidates. This loop handles the stepwise addition of as many subsequent predictors as desired.

```
while (nkept < maxkept) {          // While still adding predictors

  bestcrit = -1.e60;               // Will be criterion of the best candidate
  for (icand=0; icand<n_indep_vars; icand++) { // Try all candidates
    for (i=0; i<nkept; i++) {      // Is this candidate already in kept set?
      if (kept[i] == icand)        // If it's there
        break;                     // Quit searching for it
    }
    if (i < nkept)                 // If this candidate 'icand' is already kept
      continue;                    // Skip it

// Compute I(Y;X|Z) for each Z in the kept set, and keep track of min
// We've already done them through last_indices[icand], so start
// with the next one up. Allow for early exit if icand already loses.

    for (iz=last_indices[icand]+1; iz<nkept; iz++) { // Continue checking all Zs
      if (scores[icand] <= bestcrit)   // Has this candidate already lost?
        break;                         // If so, no need to keep doing Zs
      j = kept[iz];                    // Index of variable in the kept set
      temp = mutinf_b (ncases, bins_dep, bins_indep + icand * ncases,
                  bins_indep + j * ncases); // I(Y;X|Z)
      if (temp < scores[icand])        // Keep track of min across all Zs
        scores[icand] = temp;
      last_indices[icand] = iz;        // Also remember how far we've checked
    } // For all kept variables, computing min conditional mutual information
```

```
      criterion = scores[icand];        // Equation (2.19), possibly abbreviated
      if (criterion > bestcrit) {        // Did we just set a new record?
         bestcrit = criterion;           // If so, update the record
         ibest = icand;                  // Keep track of the winning candidate
         }
      } // For all candidates

   // We now have the best candidate
   kept[nkept] = ibest;
   crits[nkept] = bestcrit;
   ++nkept;
   } // While adding new variables
```

FREL for High-Dimensionality, Small Size Datasets

The curse of data miners is the situation of having a large number of variables and a small dataset. If, in addition, the data is noisy, most statistical analyses are hopeless. Spurious results are virtually inevitable. Even if the data is clean, statistical analysis is difficult. But if we are looking only for relationships between a single target variable and any of a multitude of competitors, [Yun Li et al, "FREL: A Stable Feature Selection Algorithm", *IEEE Transactions on Neural Networks and Learning Systems*, July 2015.] provide an interesting algorithm called *Feature Weighting as Regularized Energy-Based Learning*, abbreviated *FREL*.

The FREL algorithm is a useful method for ranking, and even weighting, predictor candidate variables in a classification application that is relatively low noise but is plagued by high dimensionality (numerous predictor candidates) and small sample size. The implementation presented here is strongly based on their innovative algorithm, but with significant modifications that I believe improve on the original version by providing more accurate and stable weights (at the cost of slower execution). My implementation also includes an approximate Monte Carlo permutation test (MCPT) of the null hypothesis that all predictors have equal value, as well as an MCPT of the null hypothesis that the predictors, taken as a group, are worthless. Sadly, I am unable to devise a FREL-based MCPT of any null hypothesis concerning individual predictors taken in isolation. We'll discuss these issues in more detail later.

The next three or four pages will present a fairly theoretical discussion of the FREL algorithm in its most general form. Feel free to skim them. Understanding the theory is not necessary to program and use FREL.

141

The model that inspires FREL is weighted nearest-neighbor classification. The distance between a test case having predictors $x = \{x_1,..., x_K\}$ and a training-set case $t = \{t_1, ..., t_K\}$ is defined as the city-block distance between these cases, with each dimension having its own weight. This is defined in Equation (2.21).

$$D(x,t) = \sum_k w_k |x_k - t_k| \qquad (2.21)$$

Then, if we want to classify an unknown test case x based on a training set, we would compute the distance between the test case and each member of the training set. The chosen class for the test case would be the class of the training case having minimum distance from the test case.

Of course, performing this classification presupposes that we know appropriate weights. The procedure can be inverted and used to find optimal weights, and we could then interpret the weights as measures of importance of the predictors (assuming that the predictors have commensurate scaling!). All we would do is define a measure of classification quality and then find weights that maximize this quality measure.

An approach to machine learning that is becoming more and more popular is energy-based modeling. We have a set of random variables, which in the current context would be predictors, and a prediction target or class membership. The model defines a scalar energy as a function of the values of these variables, sometimes called their configuration. This energy is a measure of the compatibility of the configuration, with small values of energy corresponding to compatible configurations. If we have a known energy-based model and we want to make an inference (a prediction or classification) based on specified values of the predictors, we fix the predictors and vary the target or class variable to identify the configuration that minimizes the energy.

To find a good energy-based model, we tune the parameters of the model in such a way that "correct" configurations (as indicated by the training set) have small energy and "incorrect" configurations have large energy.

Once the structure of the model is specified, to find optimal parameters we define a loss functional (a function of a function). The model is a function that maps configurations of variables to energy values, and the loss functional maps models to scalar loss values. To train the model, we find the version (parameters for the model family) that minimizes the loss functional.

The most common version of this latter operation, which we will do here, is to define a per-sample loss functional as a function of the model and a single case and then average this per-sample measure across the entire training set.

This is a good time for a brief digression to make sure that two crucial issues are clear. First, many models, such as nearest-neighbor classification and some types of kernel regression, implicitly include the entire training set (or some other dataset) as a key component of the model. Do not confuse this with discussions of the training set related to training. It's still just the model, and we need not explicitly mention the presence of the training set as part of the model. Any "training set" that is an essential component of the model and the training set that we are using for optimizing the model are conceptually different entities, which may or may not actually be the same data. We simply ignore any "training set" that happens to be part of the model. Just think about the model.

Second, do not confuse energy with loss. *Energy* is a measure of the compatibility of a given variable configuration with a model, and it *is used to make a prediction. Loss* is a measure of the quality of a model in a way that generally is based on a training set, and it *is used to find an optimal model.*

The energy that a model M assigns to a hypothetical variable configuration $\{x, y\}$ can be conveniently written as $E(M, x, y)$. An extremely common and useful way to express the per-sample loss for a single training case $\{x^i, y^i\}$ is $L(y^i, E(M, x^i, \Upsilon)$, in which the term $E(M, x^i, \Upsilon)$ actually stands for multiple energy values, one for each possible value of y. In other words, the per-sample loss for a single case is a function of the true value of y for that case, and the energies given by the model for x associated with every possible y.

Note, by the way, that the distinction between function and functional become a bit murky here, depending on whether we think in terms of E being an observed number or a hypothetical function. In any case, the idea should be clear from context.

We are almost done presenting a general form of an effective loss function(al) for training an optimal (in the sense of the loss) model. We have seen the form of a per-sample loss and stated that averaging this quantity over every sample in the training set is reasonable. The only remaining issue is that of regularization. This enables us to embed prior knowledge about the model in the final solution. Typically, this involves limiting the size of weights involved in the expression of the model, although other approaches are possible. With these things in mind, we can express the loss of a given model M for a given training set T (K cases) and regularization function R as shown in Equation (2.22). This is a scalar quantity that we will minimize in order to develop a good model.

$$L(M,T) = \frac{1}{K}\sum_{k} L\left[y^k, E\left(M, x^k, \Upsilon\right)\right] + R(M) \qquad (2.22)$$

To review, a good model will fulfill two requirements: it will have low energy for correct configurations and high energy for incorrect configurations. Looked at another way, when a good model is presented with a set of predictors x, its energy will be low when it is simultaneously presented with the correct y for that x, and its energy will be high when it is simultaneously presented with any incorrect y.

It is tempting, and often appropriate, to consider only the first half of this two-part requirement: the model will have low energy for correct configurations. This is especially true for models in which fulfilling the first half automatically fulfills the second half. As an example of this situation, suppose we have a regression equation as the model, and we define the energy associated with the model and a training case as the squared difference between the correct answer and the answer provided by the regression function. If we define the loss as this energy, then averaged across the entire training set, the loss is the mean squared error (MSE). The optimal model is produced by minimizing the MSE, a venerable approach.

The regression model just used as an example is a simple, common situation. But for many model architectures, this halfway method is not a good approach. It is much better, if not mandatory, to explicitly take into account the second half of the requirement: the energy of incorrect answers should be large. And intuitively, we don't much care about easy situations, which are those incorrect answers that have huge energy. Even a weak model will do well with them. *What we must worry about is those situations in which an incorrect answer has dangerously low energy.* We want our model to be able to raise the energy of these problematic cases as much as possible above the energy of the correct answer.

This intuition leads to the following definition: The ***most offending incorrect answer*** for a case, which we will call $\ddot{y}$, is the incorrect answer that has the lowest energy. This is the answer most likely to cause an error because it is the incorrect answer that is most difficult for the model to distinguish from the correct answer. The second half of the training criterion discussed earlier, that incorrect answers should have large energy, is more general than is necessary. All we really care about is that the most offending incorrect answer has energy as large as possible, compared to the energy of the correct answer. The other incorrect answers are of lesser importance because they are easier for the model to avoid.

In particular, what we often want to maximize is the difference between the energy of the most offending incorrect answer and the energy of the correct answer. This will give us a model that is optimal in the sense of effectively handling the most difficult cases, while letting the easy cases slide.

A popular per-sample loss criterion, and which is presented here, is the log loss shown in Equation (2.23). Note how it is a monotonic function of the difference between the two energies, so optimizing either is equivalent to optimizing the other (for a single case i, not averaged across the training set!).

$$Loss\left(M,x^i,y^i\right)=\log\left(1+\exp\left[E\left(M,x^i,y^i\right)-E\left(M,x^i,\ddot{y}^i\right)\right]\right) \qquad (2.23)$$

Now that a theoretical foundation is laid, we can apply these ideas to the specific model used in the FREL paper and this text. Recall from the beginning of this section that we use weighted nearest-neighbor classification. Thus, in order to compute $E(M, x^i, y^i)$ for training case i, we check all other training cases in the correct class, y^i. The smallest distance is the energy for the correct class. Similarly, to compute $E(M, x^i, \ddot{y}^i)$, we search all other training cases in an incorrect class and find the distance to the nearest. Of course, although this is simple to describe and implement, it can be horrendously slow to compute. The quantity being minimized is the average across the training set of the per-sample losses shown in Equation (2.23). If there are n training cases and K predictors, a single evaluation of the grand loss function requires on the order of Kn^2 operations. Yikes! Luckily, FREL is most useful for situations in which the training set is small relative to the number of predictor candidates, so that squared term will ideally not be a serious problem.

Regularization

All that remains to be settled is the regularization. In any reasonable application, the energy of the incorrect answers will, on average, exceed that of the correct answers; otherwise, the model would be worthless! For the loss function shown in Equation (2.23) applied to weighted nearest-neighbor classification, increasing the weights together will decrease the loss because the term being exponentiated will become increasingly negative. Thus, naive minimization of the loss will result in the weights blowing up without bound. Thus, we are inspired to penalize large weights. This is common practice, even in situations in which this blowup is not natural. The reason is that in many models, large weights are associated with overfitting and poor out-of-sample performance. Here we use the common method of penalizing by the sum of the squares of the weights, multiplied by a user-specified regularization factor. The sum of their absolute values is also common and may be implemented easily if desired.

As we will see on page 151 when the FREL code is presented, I implement a separate weight stabilization scheme that kicks in if weights grow unreasonably large. If the user sets a positive regularization factor, this scheme will almost never play a role in optimization. However, if the user does not call for regularization (factor is zero), this scheme will prevent unrestrained runaway. For this reason, the regularization factor in my algorithm is a fairly noncritical parameter.

In practical terms, the effect of the regularization factor is to control the relative spread of weights. Suppose that predictability is concentrated in just one or a few candidates. If the user specifies a small or zero value for this parameter, the computed weights will strongly reflect this focus. However, if a large regularization factor is specified, the focus will be less intense; some of the weight will be redistributed away from the dominant predictors and given to predictors of lesser value. Intense focus on one or a few dominant predictors can, in some cases, be seen as a form of overfitting, but in other cases it is simply the "correct" response to the situation. I recommend that the user try several degrees of regularization (in *any* modeling scheme!) and compare results.

Interpreting Weights

The optimal weights determined by minimizing (possibly regularized) loss can be interpreted as measures of importance of the individual predictors. However, two issues must be considered. First, the scaling of the predictors obviously impacts the weights, so their scaling should be commensurate. In my code, I take care of this by automatically scaling per their standard deviation, though some users may want to do it differently or not at all. Second, interpretation by the user is aided by normalizing the weights in some way for display. In this presentation, they are linearly normalized so as to sum to 100.

Bootstrapping FREL

A frequently useful variation on the naive algorithm described so far is to take many bootstrap samples from the dataset and compute the final weight estimate by averaging the estimates produced from each bootstrap sample. The sampling must be done without replacement, as nearest-neighbor algorithms are irreparably damaged when the dataset contains exact replications of cases. Bootstrapping FREL has at least two major advantages over doing one FREL analysis of the entire dataset.

- Stability is usually improved. A critical aspect of any weighting scheme is that the computed optimal weights should be affected as little as possible by small changes in the dataset. Such changes might be inclusion or exclusion of a few training cases or the addition of noise to the data. An average of bootstraps is much more robust against data changes compared to a single complete FREL processing.

- Because run time of the FREL algorithm is proportional to the square of the number of cases, we can greatly decrease the run time by performing many iterations of a small sample.

For these reasons, bootstrapping is generally recommended. The sample size must be large enough that each sample is virtually guaranteed to have a significant number of representatives from each target class. For the number of iterations, my own rough rule of thumb is that the product of the number of iterations times the sample size should be about twice the number of training cases.

Monte Carlo Permutation Tests of FREL

A Monte Carlo permutation test is a useful, though time-consuming, way to test certain null hypotheses about the predictor candidates subjected to the FREL algorithm. It is vital to understand that these tests are significantly different from the permutation tests described starting on page 89. For one thing, I am not aware of any way of performing a perfect *individual-candidate* MCPT with FREL; the best I can do is come up with a rough approximation that appears to work well in practice. In the univariate screening tests described previously, the candidate predictors are handled individually, so the p-values (at least the solo tests) are independent. But FREL considers all candidates simultaneously. This dependence changes the nature of the MCPT. One effect is for dominant candidates to "suck" weight out of lesser candidates, thus reducing their apparent significance. But the most important effect is to radically change the nature of the null and alternative hypotheses of the test.

In univariate screening tests, the null hypothesis for each solo p-value is that the individual candidate is worthless, and the null hypothesis for the unbiased p-values is that all candidates are worthless. The power of the test is in identifying individual candidates that have predictive power. But for FREL, the individual MCPT tests have no useful power in situations in which all candidates have equal predictive power, regardless of whether that power is tiny or large. The null hypothesis is still generated by making all candidates worthless, exactly as in other tests. But because of the joint estimation of weights, it is more intuitive (though not strictly correct!) to think of the null hypothesis as being that all candidates have *equal* predictive power, with the unbiased p-values compensating for the fact that we are testing numerous candidates, and any of them may be outstanding by random luck. In other words, these individual tests are related to the predictive power of each candidate *relative to their competitors*. Their individual predictive powers play no easily identifiable role in determining p-values.

With this in mind, we can look at the p-values of candidates at the top of the list, those ranked highest in terms of predictive power and having the largest weights, and consider the p-values as being the probability that if all candidates were truly equal in predictive power, the top-ranked candidates would have outperformed the others to the degree shown. Suppose we see a highly significant result for the single best candidate. It may be that this best candidate is *almost* worthless, and its competitors are *completely* worthless. Or it may be that this single candidate is excellent, while its competitors are merely very, very good. In either case we may see the best candidate having a highly significant p-value. *We don't know which situation is true; it's all relative.* Again, I emphasize that this interpretation is not strictly correct, but I believe that it is close enough, especially the unbiased p-values, to be effective indicators of the validity of the obtained results.

The sucking of weight from relatively poor predictors to good predictors has a peculiar and potentially confusing effect on the solo p-values. As we drop down the sorted list to the low-ranked candidates, we can see the solo p-values cover a wide range, jumping up and down between high and low significance randomly. This is illustrating in an exaggerated manner the fact that the p-values for worthless candidates in any statistical test have a uniform distribution, with all values being equally likely. This is yet another reason why we should focus on the unbiased p-values, ignoring the solo p-values except perhaps (and with great caution) for the few top-ranked candidates.

We can compute one additional p-value, which I call the *Loss p-value*. This is a "grand" measure of the ability of all predictors taken together to be effective at correct classification. The null hypothesis is that none of the candidates has any predictive power, and the Loss p-value is the probability that if this were so, we would have achieved a loss at least as low (good) as that obtained. *This p-value being small is a necessary condition for any of the individual p-values to be meaningful.* If we cannot be reasonably certain that at least one of the candidates has predictive power, then there is no point in considering their relative power!

General Statement of the FREL Algorithm

In the next section we'll explore an efficient C++ implementation of the FREL algorithm. However, if you want to program it in a different language and want just a general outline, as well as to help C++ programmers understand the relative complex code that follows, I'll first present my implementation of the FREL algorithm in its most general form, avoiding language-specific code as much as possible. In keeping with common practice when stating algorithms, we'll use origin-one subscripting, even though C++ uses origin zero.

We begin with the core routine that is given a set of cases (predictor competitor matrix and target class vector) and a trial weight set. It computes the loss associated with this dataset and weight set. Here is the algorithm, and comments follow:

```
Subroutine compute_loss (Ncases, PredictorVecs, ClassVec, Weights)

loss = 0
For outer_case from 1 to Ncases
  ebest = eworst = infinite

  For inner_case from 1 to Ncases
    If inner_case == outer_case
      continue
    Use Eq 2.21 on Pg 142 to compute distance between inner_case and outer_case
    If ClassVec[inner_case] == ClassVec[outer_case]
      If distance < ebest
        ebest = distance
    else
      If distance < eworst
        eworst = distance
    End of inner_case loop

  loss += log (1.0 + exp (ebest - eworst)) Equation (2.23) on Page 145
  End of outer_case loop

loss += regularization penalty  Complete Equation (2.22) on Page 143
Return loss
```

The outer_case loop will cumulate the sum of Equation (2.22) on page 143. Look back at Equation (2.23) on page 145. We'll use an inner loop that checks every training case except the one being tested. At the end of this checking, we'll have the first term of Equation (2.23), the energy of the correct answer, in ebest. Also, we'll have the second term, the energy of the *most offending incorrect answer*, in eworst. The loss computed with Equation (2.23) is summed per Equation (2.22). After the sum is complete across the entire training set, we add in any desired regularization penalty.

We now present the routine that estimates the weights by combining bootstrap samples and calling an optimization routine. We'll need a subroutine that, given a set of predictors and the target class vector, finds the optimal weights, which are those that minimize the loss as computed by compute_loss(). I find that Powell's algorithm, implemented in POWELL.CPP, does a respectable job. Feel free to use a different optimizer if you want. Here is the bootstrapped weight estimator; a brief discussion follows:

Subroutine compute_weights ()

total_loss = 0
For i from 1 to Npredictors
  TotalWeights[i] = 0

For iboot from 1 to Nbootstraps
  Select BootSize cases from complete training set without replacement
  Call optimizer with these cases to find weights which minimize compute_loss()

  total_loss += this minimized loss
  For ivar from 1 to Npredictors
    TotalWeights[ivar] += OptimalWeights[ivar]
    End of ivar loop
  End of iboot loop

For ivar from 1 to Npredictors
  TotalWeights[ivar] /= Nbootstraps
  End of ivar loop

Return total_loss

This routine cumulates the total loss for all bootstrap samples. This quantity has only one use: computation of the MCPT *Loss p-value* discussed at the end of the section that begins on page 147. This lets us test the null hypothesis that *all* predictor candidates are worthless versus the alternative that at least one of the competitors has predictive power.

We estimate the weight for each candidate predictor by taking Nbootstraps samples of size BootSize, without replacement, from the complete dataset. The optimal weights for each bootstrap sample are summed, and then the sum is divided by the number of bootstraps in order to get an average. This was discussed on page 146.

At last we can present the overall FREL procedure, including the Monte Carlo permutation tests. Here is a general statement of the algorithm:

```
For irep from 1 to MCPTreps
  if irep > 1
    Shuffle target

  this_rep_loss = compute_weights()

  sum = 0
  For ivar from 1 to Npredictors
    weights[ivar] *= standard_deviation[ivar]
    sum += weights[ivar]
    End of ivar loop

  For ivar from 1 to Npredictors
    weights[ivar] *= 100 / sum
    End of ivar loop

  For ivar from 1 to Npredictors

    if (ivar == 1 || weights[ivar] > best_crit)
      best_crit = weights[ivar];

    if (irep == 1) {      // Original, unpermuted data
      original_weights[ivar] = weights[ivar]   // Save unpermuted weights
      mcpt_bestof[ivar] = mcpt_solo[ivar] = 1;
      }
```

```
    else if (weights[ivar] >= original_weights[ivar])
      ++mcpt_solo[ivar];

    End of ivar loop

  if (irep == 1)     // Original, unpermuted data
    original_loss = this_rep_loss;
    mcpt_loss = 1;

  else
    if (this_reploss <= original_loss)
      ++mcpt_loss;
    For ivar from 1 to Npredictors
      if (best_crit >= original_weights[ivar])
        ++mcpt_bestof[ivar];
      End of ivar loop

  End of irep loop

For ivar from 1 to Npredictors
  mcpt_solo[ivar] /= MCPTreps
  mcpt_bestof[ivar] /= MCPTreps

mcpt_loss /= MCPTreps
```

The main loop performs the MCPT replications. Remember that in this outline, we use origin-one to conform to common standards, with the first (unpermuted) replication being irep=1. In the C++ code that you'll see later, the origin is zero.

If we are past the first replication, shuffle the target class vector. Then compute the optimal weights for the candidate predictors.

The next two blocks of code normalize the weights. Multiplying each weight by the standard deviation of the corresponding predictor makes the resulting weights independent of scaling, which is what we want in most applications. Keep in mind that a prudent user will not rely on this operation and instead will make sure that the predictors are commensurately scaled in advance. Significant differences in scaling degrade performance of the optimizer. Then, each weight is divided by their sum and multiplied by 100. This produces weights that sum to 100, an aid to interpretability.

The next loop, which covers each predictor, does three things. First, it keeps track of the best performer's criterion, best_crit, which will soon be needed. Second, if this is the first (unpermuted) replication, it saves the "true" weights and initializes the weight MCPT counters. Third, if this is a shuffled replication, it updates the solo MCPT counters.

After this loop is finished, we will have the best criterion in best_crit. We also have the loss for this replication in this_rep_loss. If this is the first, unpermuted replication, save this loss and initialize the MCPT loss counter. Otherwise, update this counter. Then, for each predictor candidate, compare the best criterion to that predictor's original criterion in order to implement the unbiased test. Recall that strictly speaking, this test is not valid for any predictor other than the best. But as discussed earlier, these p-values are of some interest.

When all MCPT replications are complete, divide the counters by the number of replications to get the estimated p-values. If these actions are not clear, please review the MCPT section that begins on page 89, as well as the specialized FREL issues that are discussed on page 147.

Multithreaded Code for FREL

The prior section discussed the FREL algorithm in general terms. Now we will dig into specifics, especially focusing on how the potentially slow FREL algorithm can be multithreaded to take advantage of modern processors. This code is extracted from FREL.TXT.

We begin with the core routine, which corresponds to the compute_loss() algorithm shown on page 149. The overwhelming fraction of total FREL compute time is spent in the innermost (ivar) loop of this routine, so every effort should be made to make it as efficient as possible.

Here is the calling parameter list. Because the work will be split across threads, we specify starting and stopping indices of cases being tested. The indices array identifies the ncases cases in this bootstrap sample taken from the complete database. Each element in this array is a row number in the database. The database can contain more variables (columns) than the npred predictors being tested, so preds identifies the variables (columns in database) we want to test. Note that if we were not multithreading, ncases would equal istop minus istart.

```
static double block_loss (
   int istart,              // Index of first case being tested
   int istop,               // And one past last case
   int *indices,            // Index of cases; facilitates bootstraps
   int npred,               // Number of predictors
   int *preds,              // Their column indices in 'database' are here
   int ncases,              // N of cases in this bootstrap
   int n_vars,              // Number of columns in database
   double *database,        // Full database, ncases rows and n_vars columns
   int *target_bin,         // Ncases vector of target bin indices
   double *weights          // Input of weight vector being tried
   )
{
   int k, ivar, icase, inner, iclass, inner_index, outer_index;
   double *cptr, *tptr, distance, ebest, eworst, loss;
```

There are three nested loops. The outermost determines the case being tested, and this is the dimension that is split across threads. The middle loop passes across the entire sample except for the case being tested, finding the two E terms in Equation (2.23) on page 145. The innermost loop computes the city-block distance, Equation (2.21) on page 142. It may help to study the compute_loss() algorithm shown on page 149 in conjunction with this listing.

```
loss = 0.0;
for (icase=istart; icase<istop; icase++) {
   outer_index = indices[icase];          // Index of this case in complete database
   iclass = target_bin[outer_index];      // Its class
   cptr = database + outer_index * n_vars; // Its predictors in database
   ebest = eworst = 1.e60;

   // Find the two E terms in Equation (2.23) on Page 145
   for (inner=0; inner<ncases; inner++) { // Test against all other cases
      inner_index = indices[inner];       // Index of this case in complete database
      if (inner_index == outer_index)     // Don't test it against itself
         continue;
      tptr = database + inner_index * n_vars; // Predictors of inner case in database
```

```
    // Compute the distance of this inner case from the test case
    distance = 0.0;
    for (ivar=0; ivar<npred; ivar++) {    // For all predictors
       k = preds[ivar];                   // Index of this predictor in database
       distance += weights[ivar] * fabs (cptr[k] - tptr[k]); // Eq 2.21 on Page 142
       }

    // Find the closest neighbor in this class and in any other class
    if (target_bin[inner_index] == iclass) {
       if (distance < ebest)
          ebest = distance;
       }

    else {
       if (distance < eworst)
          eworst = distance;
       }
    } // For inner, the test cases

    distance = ebest - eworst;

    // Sum Equation (2.22) on Page 143
    if (distance > 30.0)      // Prevent overflow. This is harmless.
       loss += distance;
    else
       loss += log (1.0 + exp (distance)); // Equation 2.23 on Page 145

    } // For icase

  return loss;
}
```

Note that the loss function, Equation (2.23) on page 145, must not be allowed to overflow when exponentiating. So we test it against 30, and substitute an essentially equal value if we are approaching overflow.

As is standard in my work, we define a data structure for passing parameters and use a wrapper function that is executed in the threads.

```
typedef struct {
  int istart;          // Index of first case being tested
  int istop;           // And one past last case
  int *indices;        // Index of cases; facilitates bootstraps
  int npred;           // Number of predictors
  int *preds;          // Their indices are here
  int ncases;          // Number of cases in this bootstrap
  int n_vars;          // Number of columns in database
  double *database;    // Full database
  int *target_bin;     // Bin index for targets
  double *weights;     // Weight vector
  double *loss;        // Computed loss function value is returned here
} FREL_PARAMS;

static unsigned int__stdcall block_loss_threaded (LPVOID dp)
{
  *(((FREL_PARAMS *) dp)->loss) = block_loss (((FREL_PARAMS *) dp)->istart,
                  ((FREL_PARAMS *) dp)->istop,
                  ((FREL_PARAMS *) dp)->indices,
                  ((FREL_PARAMS *) dp)->npred,
                  ((FREL_PARAMS *) dp)->preds,
                  ((FREL_PARAMS *) dp)->ncases,
                  ((FREL_PARAMS *) dp)->n_vars,
                  ((FREL_PARAMS *) dp)->database,
                  ((FREL_PARAMS *) dp)->target_bin,
                  ((FREL_PARAMS *) dp)->weights);

  return 0;
}
```

The following routine splits the work across multiple threads. Blocks of code will be interspersed with discussions. The calling parameter list contains many items already discussed, so we dispense with redundant explanations.

```
static double loss (
   int npred,              // Number of predictors
   int *preds,             // Their indices (columns in database) are here
   int ncases,             // Number of cases in this bootstrap
   int n_vars,             // Number of columns in database
   int *indices,           // Index of cases; facilitates bootstraps
   double *database,       // Full database
   int *target_bin,        // Ncases vector of target bin indices
   double *weights,        // Input of weight vector being tried
   double regfac           // Regularization factor
   )
{
   int i, ivar, ithread, n_threads, n_in_batch, n_done, istart, istop, ret_val;
   double loss[MAX_THREADS], total_loss;
   FREL_PARAMS frel_params[MAX_THREADS];
   HANDLE threads[MAX_THREADS];

   n_threads = MAX_THREADS;
   if (n_threads > ncases)         // No sense multithreading a tiny problem
      n_threads = 1;

/*
   Initialize those thread parameters which are constant for all threads.
*/

   for (ithread=0; ithread<n_threads; ithread++) {
      frel_params[ithread].npred = npred;
      frel_params[ithread].preds = preds;
      frel_params[ithread].ncases = ncases;
      frel_params[ithread].n_vars = n_vars;
      frel_params[ithread].indices = indices;
      frel_params[ithread].database = database;
      frel_params[ithread].target_bin = target_bin;
      frel_params[ithread].weights = weights;
      frel_params[ithread].loss = &loss[ithread];
      } // For all threads, initializing constant stuff
```

157

```
istart = 0;        // Batch start = training data start
n_done = 0;        // Number of training cases done so far

for (ithread=0; ithread<n_threads; ithread++) { // Will launch all threads at once
   n_in_batch = (ncases - n_done) / (n_threads - ithread); // Cases left / batches left
   istop = istart + n_in_batch;                 // Stop just before this index

   // Set the pointers that vary with the batch: the starting and stopping cases

   frel_params[ithread].istart = istart;
   frel_params[ithread].istop = istop;
   threads[ithread] = (HANDLE) _beginthreadex (NULL, 0, block_loss_threaded,
                                 &frel_param s[ithread], 0, NULL);
   n_done += n_in_batch;      // Count how many cases done so far
   istart = istop;            // Start the next batch right after last case in this one
   } // For all threads / batches
```

At this point, all data has been launched, split across n_threads threads. Now we just sit and wait for them to finish. Note that error handling is omitted here for clarity. You can find it in FREL.TXT.

```
WaitForMultipleObjects (n_threads, threads, TRUE, 1200000);
```

The summation across all training cases in this bootstrap sample, each being used as a test case, was split across multiple threads. We sum the results for the threads to get the total loss for this bootstrap sample. Also, close the threads so as to be a responsible and thrifty Windows user. Last of all, add in the regularization penalty.

```
total_loss = 0.0;
for (ithread=0; ithread<n_threads; ithread++) {
   total_loss += loss[ithread];
   CloseHandle (threads[ithread]);
   }

total_loss /= ncases;  // Make it a per-case average

// Add in the regularization penalty
for (ivar=0; ivar<npred; ivar++)
   total_loss += regfac * weights[ivar] * weights[ivar];

return total_loss;
}
```

We come now to the code that does the bootstrap sampling and repeatedly call the loss() function just presented, pooling the bootstrapped weight estimates and loss. The calling parameter list is shown on the next page. But we begin with a bunch of static declarations. These are a sneaky but efficient way of passing parameters to the criterion routine that will be called by the optimizer. By doing it this way, we can use a general-purpose optimization routine, avoiding the need for a routine specialized for this particular application.

```
static int criter (double *x, double *y);   // Computes the criterion being minimized
static int local_npred;                      // These are the same parameters that
static int *local_preds;                     // we've been seeing in prior routines
static int local_ncases;                     // As before, this is the bootstrap sample size
static int local_n_vars;
static int *local_indices;
static double *local_database;               // The entire database, all trainng cases
static int *local_target_bin;
static double *local_critwork;
static double local_regfac;
static int compute_wt (
    int npred,            // Number of predictors
    int *preds,           // Their indices are here
    int ncases,           // Number of cases in complete database
    int n_vars,           // Number of columns in database
    int *indices,         // Index of cases; facilitates bootstraps
    double *database,     // Full database
    int nbins_target,     // Number of target bins
    int *target_bin,      // Ncases vector of target bin indices
    int nboot,            // Number of bootstrap reps
    int bootsize,         // Size of each bootstrap
    double *crits,        // Predictor weights for each bootstrap computed here
    double *critwork,     // Work vector npred long needed by criter()
    double *base,         // Work vector npred long for powell()
    double *p0,           // Work vector npred long for powell()
    double *direc,        // Work vector npred*npred long for powell()
```

```
double regfac,          // Regularization factor
double *loss_value,     // Optimal loss (sum of bootstrap losses) is returned here
double *weights         // Weight vector returned here
)
{
int i, j, k, m, iboot, ret_val, class_count[MAX_MUTINF_BINS];
double loss;
char msg[2014];

// These are needed by criter()

local_npred = npred;
local_preds = preds;
local_ncases = bootsize;
local_n_vars = n_vars;
local_indices = indices;
local_database = database;
local_target_bin = target_bin;
local_critwork = critwork;
local_regfac = regfac;
```

We do a few things to initialize for the bootstrapping. The final weights will be the mean weight estimates across all bootstraps. We'll also sum the loss across all bootstraps, which will be used only for a particular MCPT described later. Finally, we initialize the vector that will specify the case indices for each bootstrap replication.

```
for (i=0; i<npred; i++)   // Results of bootstraps will be summed in 'weights'
   weights[i] = 0.0;

*loss_value = 0.0;        // Will be needed for global p-value

for (i=0; i<ncases; i++)
   indices[i] = i;        // Identifies cases in each bootstrap sample
```

Here is the bootstrap loop. Because we use a nearest-neighbor algorithm as part of the criterion calculation, no case can be replicated in the sample. The easiest way to select without replacement is to shuffle in place and stop when we reach the bootstrap size. The first bootsize cases in the shuffled array define the bootstrap sample. We'll discuss this code in a moment.

```
for (iboot=0; iboot<nboot; iboot++) {

  for (i=0; i<nbins_target; i++)
    class_count[i] = 0;   // This will be used in the next section of code

  i = ncases;             // Number remaining to be shuffled
  while (i > 1) {         // While at least 2 left to shuffle
    m = ncases - i;       // Number shuffled so far
    if (m >= bootsize)
      break;
    j = (int) (unifrand_fast () * i);
    if (j >= i)           // Should never happen, but be safe
      j = i - 1;
    k = indices[m];
    indices[m] = indices[m+j];
    indices[m+j] = k;
    --i;
    ++class_count[target_bin[indices[m]]];     // We'll need this in a moment
  } // Shuffling for bootstrap sample without replication
```

The first action in the bootstrap loop is to initialize every element of class_count to
zero. These will count the number of occurrences of each class in the sample. You'll
learn more about this soon.

The shuffling loop shown previously is similar to the standard algorithm but
changed so that shuffling moves from beginning to end instead of the more common
end to beginning. That would have worked as well, but it's more intuitive to submit the
beginning of the array as the bootstrap rather than the end. That's just my opinion.

To make sure this technique is clear, we'll explore its actions. The counter i will always
be the number of elements in the indices array that are not yet shuffled. It is initialized to the
number of cases in the complete database. Then m = ncases - i is the number that have been
shuffled, all of which will be at the beginning of the array. If we have reached the required
number of cases (bootsize) for this sample, we are done. If not, we choose j randomly from
the number of as-yet unshuffled cases. Fetch this randomly selected case and put in the
next spot, swapping what was there into the slot from which we just fetched a case. This
way, every case in the bootstrap sample will have an equal chance of being any dataset case
except for any case that has already been selected for the sample. We also update the counter
of how many times each target class has appeared in this bootstrap sample.

161

The weight estimation algorithm will misbehave if we have no cases in some class. I set an arbitrary limit of requiring at least two cases in each class. If this requirement is not met, we reject this sample and try again.

```
for (i=0; i<nbins_target; i++) { // Demand at least two of each class in this sample
  if (class_count[i] < 2)
    break;
  }

if (i < nbins_target) {
  --iboot;
  continue;
  }
```

The rest of this routine is fairly simple. As we'll see in the next module, rather than optimizing the weights themselves, we optimize the log of the weights. This aids numerical stability. So we initialize the starting point for optimization to zero, which corresponds to weights of one. The powell() minimization routine requires that we provide the function value (the loss here) at the starting point, so we call the criterion function to get this quantity and then call the optimizer. Cumulate across bootstraps the loss and the optimal weights. Finally, after all bootstraps are complete, divide the sum of weight estimates by the number of bootstraps to get their average.

```
for (i=0; i<npred; i++) // Starting point for this bootstrap
  crits[i] = 0.0;

ret_val = criter (crits, &loss);
ret_val = powell (0.1, 10, 0.0, 1.e-3, criter, npred,
          crits, &loss, base, p0, direc, 1);
*loss_value += loss;

for (i=0; i<npred; i++) // Cumulate for this bootstrap
  weights[i] += crits[i];

  } // For iboot

for (i=0; i<npred; i++)
  weights[i] /= nboot;
}
```

We won't bother discussing the Powell's method optimizer here; it is well documented in numerous references. The code for it is supplied in POWELL.CPP. You should feel free to substitute your own optimizer if you have something you think is better. Also feel free to tweak the convergence parameters in this function call. See POWELL.CPP for details.

What about this criter() routine that, given a trial set of weights, computes the loss for the current bootstrap sample? Here is the code, and a brief explanation follows:

```
static int criter (double *x, double *y)
{
  int i;
  double crit, penalty;

  penalty = 0.0; // This is not regularization. It just keeps the parameters reasonable.

  for (i=0; i<local_npred; i++) {
    if (x[i] > 4.0) {
      local_critwork[i] = exp (4.0) + x[i] - 4.0;
      penalty += (x[i] - 4.0) * (x[i] - 4.0);
      }
    else if (x[i] < -3.0) {
      local_critwork[i] = exp (-3.0) + x[i] + 3.0;
      penalty += (x[i] + 3.0) * (x[i] + 3.0);
      }
    else
      local_critwork[i] = exp (x[i]);
    }

  crit = loss (local_npred, local_preds, local_ncases, local_n_vars,
               local_indices, local_database, local_target_bin,
               local_critwork, local_regfac);

  *y = crit + penalty;

  return 0;
}
```

Regularization is done in the loss() function, not in this routine. But we do include a penalty term to prevent weight runaway, which will almost never be invoked if even slight regularization is done. Recall that we are optimizing the log of the weights. If this log grows too large (> 4) or small (< -3), we modify the variable-to-weight mapping function in a way that does not introduce discontinuity and penalize accordingly. This is very benign and is really just cheap, innocuous insurance against bad behavior.

The hard work is done. All that remains is the main routine that calls compute_wt(), optionally with shuffling for Monte Carlo permutation testing. However, it would be wasteful to list the code in detail here, because the important concepts of this procedure were described on page 151 already. Instead, I refer the reader to the FREL.TXT file and mention a few items of interest in regard to the frel() routine and that do not appear in that earlier outline:

- This code uses the partition() routine (page 30) to group the target variable into classes. This allows maximum generality, since the target can be continuous, but if it is already discrete, the existing classes will be respected except in pathological situations.

- Full or cyclic permutation is supported.

- When the first (unpermuted) replication is performed, a copy of the weights is kept, and these are then sorted, simultaneously moving a vector of indices. This facilitates later printing of the weights in sorted order.

Some FREL Examples

Here are some simple examples of using FREL testing to evaluate the relationship of a set of competing candidates with a single target variable. The first example shows the effect of no regularization, the second demonstrates the impact of hugely excessive regularization, and the third modestly large regularization.

The synthetic variables in the dataset are as follows:

- RAND0 to RAND9 are independent (within themselves and with each other) random time series.

- SUM1234 = RAND1 + RAND2 + RAND3 + RAND4

We begin by specifying a regularization factor of zero and running 100 MCPT replications. The following results are produced:

Variable	Weight	Solo pval	Unbiased pval
RAND4	24.4017	0.0100	0.0100
RAND1	23.9127	0.0100	0.0100
RAND2	22.3636	0.0100	0.0100
RAND3	19.8841	0.0100	0.0100
RAND6	2.7574	1.0000	1.0000
RAND8	1.5689	1.0000	1.0000
RAND5	1.4971	1.0000	1.0000
RAND9	1.3692	1.0000	1.0000
RAND7	1.2613	1.0000	1.0000
RAND0	0.9839	1.0000	1.0000

Loss p-value = 0.010

Observe that the algorithm does a fabulous job of identifying the four variables that are related to the target. The weights for the good and worthless variables are very different, and both the solo and unbiased p-values could not be better.

We now use an absurdly large regularization factor, 10. As pointed out earlier, regularization tends to obscure differences between variables. We see it dramatically here, when only three of the four "good" variables make the top of the sorted list. Interestingly enough, the solo p-values still correctly identify the four good variables, while the unbiased p-values are terribly distorted. The lesson is that regularization comes at a price.

Variable	Weight	Solo pval	Unbiased pval
RAND1	10.1753	0.0100	0.0100
RAND3	10.1326	0.0100	0.0900
RAND4	10.0753	0.0100	1.0000
RAND9	10.0517	1.0000	1.0000
RAND0	10.0429	1.0000	1.0000
RAND2	9.9708	0.0100	1.0000
RAND8	9.9582	1.0000	1.0000
RAND7	9.9575	1.0000	1.0000
RAND6	9.8321	1.0000	1.0000
RAND5	9.8036	1.0000	1.0000

Loss p-value = 0.010

Finally, we use a regularization factor of 0.1, which is fairly large but not ridiculous. See how the weight difference between the "good" and the "bad" variables are uncomfortably close. Nonetheless, the p-values do an excellent job of separation.

Variable	Weight	Solo pval	Unbiased pval
RAND1	15.6745	0.0100	0.0100
RAND2	15.1372	0.0100	0.0100
RAND3	15.0183	0.0100	0.0100
RAND4	14.7490	0.0100	0.0100
RAND9	7.0528	1.0000	1.0000
RAND0	6.9595	1.0000	1.0000
RAND5	6.5893	1.0000	1.0000
RAND8	6.3851	1.0000	1.0000
RAND6	6.3514	1.0000	1.0000
RAND7	6.0830	1.0000	1.0000

Loss p-value = 0.010

CHAPTER 3

Displaying Relationship Anomalies

Naive measures of association between variables, such as linear correlation, are primarily sensitive to gross relationships, those patterns that are easy to detect, see, and describe. In prior chapters we examined measures that go beyond such naiveté and are able to detect more subtle dependencies between variables, in other words, anomalies in otherwise uncomplicated relationships. But what if we want a visual representation of the pattern that connects them? In this chapter we present several ways of doing this.

The material in this chapter, as well as many (most?) techniques for measuring relationships between variables, is based on a fundamental statistical principle: *two variables are unrelated if and only if their joint distribution equals the product of their marginal distributions.* To take a simple example from a discrete distribution, suppose Variable 1 has probability 0.3 of having value A, and Variable 2 has 0.2 probability of having value M. If these two variables are independent, the probability of simultaneously observing these values (Variable 1 = A and Variable 2 = M) is 0.3 * 0.2 = 0.06. If in an experiment we observe that for one or more pairs of outcomes, the observed joint probability is not close to the product of the observed marginal probabilities, this is evidence that the variables are not independent.

If the variables are continuous, the same rule applies, although the lack of categories makes the intuition less straightforward. Let random variables X_1 and X_2 have density functions $f_1(x_1)$ and $f_2(x_2)$, respectively. Let their joint density function be $f(x_1, x_2)$. Then X_1 and X_2 are independent if and only if $f(x_1, x_2) = f_1(x_1) f_2(x_2)$.

© Timothy Masters 2018
T. Masters, *Data Mining Algorithms in C++*, https://doi.org/10.1007/978-1-4842-3315-3_3

We can make effective use of this defining property of independence by visually displaying its components as well as deviations from equality. But a graphical display should be continuous in order to be pleasing to the eye, so we need a way of computing $f_1(x_1)$ and $f_2(x_2)$ for arbitrary values of x_1 and x_2 across their entire practical domain. We will need this ability regardless of whether the variables are discrete or continuous, and it must provide reasonable results for small samples, as well as be reasonably fast to compute for large samples. The latter requirement can be troublesome, but we'll do the best we can.

An excellent way to compute the joint and marginal densities is to use the *Parzen window* method described on page 37. You are encouraged to review that material. For convenience, the four key equations are shown here, as they will be implemented in the code that follows on page 173. Equation (3.1) is the univariate window, the ordinary exponential function, and Equation (3.2) is the corresponding univariate density estimator. Their multivariate extensions are shown in Equations (3.3) and (3.4). For our purposes, $p=2$ in these latter two equations.

$$W(d)=\frac{1}{\sqrt{2\pi}}e^{-d^2/2} \tag{3.1}$$

$$f(x)=\frac{1}{n\sigma}\sum_{i=1}^{n}W\left(\frac{x-x_i}{\sigma}\right) \tag{3.2}$$

$$W(d_1,\ldots d_p)=\frac{1}{(2\pi)^{p/2}}e^{-\frac{1}{2}\sum_{1}^{p}d_i^2} \tag{3.3}$$

$$f(x_1,\ldots,x_p)=\frac{1}{n\sigma_1\ldots\sigma_p}\sum_{i=1}^{n}W\left(\frac{x_1-x_{1,i}}{\sigma_1},\ldots,\frac{x_p-x_{p,i}}{\sigma_p}\right) \tag{3.4}$$

There are four ways of displaying these quantities that I have found useful: the marginal density product, the actual bivariate density, the marginal inconsistency, and the contribution to mutual information. We'll explore these one at a time.

To provide a simple yet revealing comparison between the four types of plot, I generated a pair of random variables, *INDEP* and *BLOB*. The former is uniformly distributed from -50 to 50. The latter is similar, except that when *INDEP* lies between 15 and 25, *BLOB* is changed to -30 plus a small uniform random variation ranging from -5 to 5. The four plots appear on the next two pages in Figure 3-1, Figure 3-2, Figure 3-3, and Figure 3-4, and explanations follow.

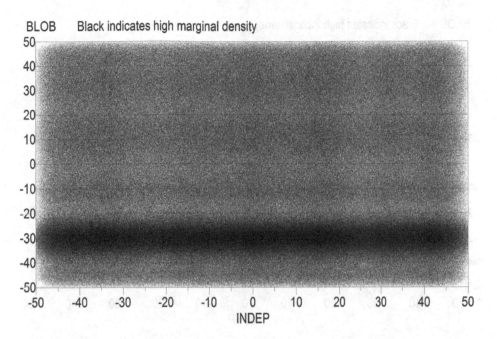

Figure 3-1. *Marginal density product*

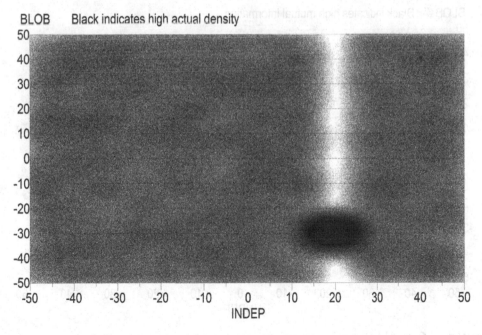

Figure 3-2. *Actual density*

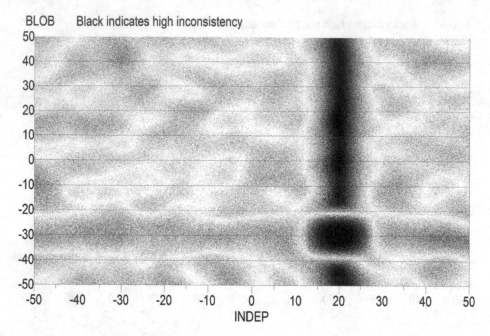

Figure 3-3. *Marginal inconsistency*

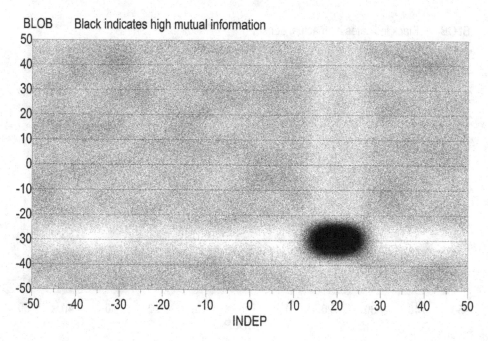

Figure 3-4. *Mutual information contribution*

Marginal Density Product

The marginal density plot shows the log of the product of the two marginal densities, $f_1(x_1) f_2(x_2)$. It is useful as a "baseline" display, as it shows the bivariate density as it would exist if there were no relationship between the horizontal and vertical variables. Of the four types of plot, this is certainly the least useful and is often worthy of being ignored.

Figure 10-1 depicts a dark horizontal band, centered in the vertical (*BLOB*) dimension at -30. It extends across the entire horizontal (*INDEP*) range. The band exists at -30 because *BLOB* cases are concentrated there. But it extends across the entire range of *INDEP* because this plot ignores any relationship between the variables. Thus, the fact that *BLOB* is shifted to -30 for only a subset of the domain of *INDEP* is of no consequence to this plot. The plot is constructed based on only the separate distributions of each variable.

Actual Density

The *actual density* plot is, in a sense, the opposite of the marginal product plot because it illustrates the full nature of the dependency between the horizontal and vertical variables. It depicts the log of the joint distribution of these two variables, $f(x_1, x_2)$. As such, one can see where cases are concentrated and where they are thinly distributed.

Figure 10-2 clearly shows how, in the 15 to 25 range of *INDEP*, values of *BLOB* are concentrated around -30. The light bands above and below this dark area show that the -30 concentration has come at the expense of other values of *BLOB* when *INDEP* is in the 15 to 25 range.

Marginal Inconsistency

Recall that two variables are independent if and only if $f(x_1, x_2) = f_1(x_1) f_2(x_2)$ everywhere. If there is even one location (x_1, x_2) where this defining property does not hold, then the variables are not independent. It is often in our interest to find those locations where this equality fails. Equation (3.5) is an effective way to measure the degree to which the joint density fails to equal the product of the marginal densities.

$$Inconsistency = ABS\left[\log\left(\frac{f(x_1, x_2)}{f(x_1) f(x_2)}\right)\right] \tag{3.5}$$

When the joint equals the marginal product, *Inconsistency* will be zero. As the two depart more and more, *Inconsistency* will increase. Sometimes it may be more useful to avoid the absolute value so that relatively sparse joint density is indicated by a negative inconsistency. However, in my own work I have found it more informative to focus on only the degree of inconsistency, regardless of sign, and use other plots to determine the nature of the inconsistency. I find that my eye responds more easily to departures from normalcy when it has to look for only one feature (abnormally positive) rather than being open to two features (abnormally positive or negative).

Figure 10-3 does an excellent job of revealing the fact that something unusual happens when *INDEP* lies in the 15 to 25 range. Density above and below the vicinity of *BLOB*=-30 gets sucked into the -30 area. Whether a region of *BLOB* is a sucker or a suckee, this inconsistent behavior in the region is flagged by large values of *Inconsistency*.

Notice the less prominent horizontal dark band around *BLOB*=-30. This is because based purely on the *BLOB* marginal, one would expect a few more cases here, but the actual joint density is too small.

Lastly, the white (low inconsistency) bands around the border of the inconsistent regions are because the Parzen window averages cases. The opposing nature of inconsistency on opposite sides of the border average out to "consistent" behavior at the border.

Mutual Information Contribution

Mutual information (page 17) is an effective measure of the degree to which two variables are related. Recall that Equation (3.6) is the fundamental definition of mutual information. The summation involves the product of two terms. One of them is the inconsistency we discussed in the prior section, though without the absolute value. The other is the probability of a potentially inconsistent location in the joint domain occurring. The summation is over the entire domain, all possible values of the two variables. It can be interesting to locate the areas of the joint domain that are the primary contributors to the mutual information.

$$I(X;Y) = \sum_{x \in \chi} \sum_{y \in \Psi} p(x,y) \log \frac{p(x,y)}{p(x)p(y)} \tag{3.6}$$

Any inconsistency between the joint density and the product of the marginals will be given weight in proportion to the probability of that region; regions in which the joint density is unusually high will be given especially large weighting of any inconsistency there.

Figure 10-4 shows this in action. The area in which cases have an unusually high concentration is prominent, a reflection of the magnitude of both terms in the product within this region. This area simultaneously has a large joint density relative to the product of the marginals (high inconsistency), and it also has an unusually high concentration of cases in this neighborhood (high actual density), thus giving large weight to the inconsistencies in this area of the domain.

The lighter vertical and horizontal bands illustrate the opposing effect: these regions have unusually low density.

Code for Computing These Plots

The file DENSITY_PLOTS.TXT contains the key computational code for generating the displayable grid for the four plots just discussed. Error checking and other aspects of the user interface have been omitted for clarity. In this section we will explore this code, section by section, to make sure its operation is clear.

The following variables will play significant roles in the code:

database	n_cases (rows) by n_vars (columns) dataset containing all data
grid	res by res displayable image which we compute
val1	Horizontal variable, which we extract from the database
val2	And vertical variable
keys	Work area, needed only for histogram equalization

The user-specified parameters are shown next. Their purposes will be explained in more detail as relevant portions of the code are presented.

varnum1	Column in the database of horizontal variable
varnum2	And vertical variable
use_lowlim1	Flag: limit the lower range of the horizontal variable?
lowlim_val1	Lower limit if specified by user
Similarly variables for upper limits and vertical variable	
res	Vertical and horizontal resolution of the square image generated
width	Fraction of standard deviation used for Parzen window width
shift	Amount to shift displayed tone for better display
spread	Amount to expand displayed tone range for better display

type	Type of display	
	TYPE_DENSITY	Actual density (similar to scatterplot)
	TYPE_MARGINAL	Marginal density, shows 'no relationship' pattern
	TYPE_INCONSISTENCY	Marginal inconsistency
	TYPE_MI	Mutual information contribution
hist	Apply histogram normalization?	
sharpen	Sharpen display range to clarify boundary?	

First, we allocate work areas. Note that if histogram normalization is not to be performed, we do not need to allocate keys. We allocate grid to be twice the display size. We'll use the second half as a scratch work area later.

```
grid = (double *) MALLOC (2 * res * res * sizeof(double));
keys = (int *) MALLOC (res * res * sizeof(int));
val1 = (double *) MALLOC (n_cases * sizeof(double));
val2 = (double *) MALLOC (n_cases * sizeof(double));
```

It's trivial to extract the data from the database. If you already have it in two arrays, you don't need to do this. From here on, we will reference val1 (the horizontal variable) and val2 (vertical) only.

```
for (i=0; i<n_cases; i++) {
  val1[i] = database[i*n_vars+varnum1];   // Horizontal variable
  val2[i] = database[i*n_vars+varnum2];   // Vertical variable
  }
```

We pass through the horizontal variable, finding the smallest and largest values, which will be used to control display scaling. If the user requests different limits for display, override the limits just found. Naturally, we could reorganize this code to avoid the loop if user-specified limits are supplied. But the loop is fast, and the code is clearer this way. Redo it if you'd like.

```
smallest = largest = val1[0];
for (i=1; i<n_cases; i++) {
  if (val1[i] < smallest)
    smallest = val1[i];
  if (val1[i] > largest)
    largest = val1[i];
  }
```

```
if (use_lowlim1)
   smallest = lowlim_val1;

if (use_highlim1)
   largest = highlim_val1;
```

A careless user may have specified conflicting limits. The following check is cheap insurance against disaster:

```
if (largest <= smallest) {    // Should never happen, but user may be careless
   largest = smallest + 0.1;
   smallest = largest - 0.2;
   }
```

At this point, the programmer would use these limits to set up labels for the display and maybe revise the display limits. Sometimes visual appearance is improved by extending the actual display limits beyond the data or user-specified limits. We leave it to you to implement this as desired. Just let (xmin, xmax) be the actual display range. Also, we perform these same operations with the vertical variable. There's no sense being redundant in this presentation.

We now compute the scale factors (sigma in the denominator of Equations (3.2) and (3.4)) for the horizontal and vertical variables. The user-specified width is the fraction of each variable's standard deviation to use for this scale factor, the width of the Parzen window.

```
scale1 = scale2 = mean1 = mean2 = 0.0;

for (i=0; i<n_cases; i++) {
   x = val1[i];
   if (use_lowlim1 && x < lowlim_val1)
      x = lowlim_val1;
   if (use_highlim1 && x > highlim_val1)
      x = highlim_val1;
   mean1 += x;
   x = val2[i];
   if (use_lowlim2 && x < lowlim_val2)
      x = lowlim_val2;
```

```
  if (use_highlim2 && x > highlim_val2)
    x = highlim_val2;
  mean2 += x;
  }

mean1 /= n_cases;
mean2 /= n_cases;
```

The previous code computes the mean of each variable, and the following code computes the standard deviation. If the user specified a display limit, we bound the variable accordingly. It can be argued that it would be better to avoid bounding when computing the mean and standard deviation. This is a personal preference. You may want to try it both ways and see which you prefer.

```
for (i=0; i<n_cases; i++) {
  x = val1[i];
  if (use_lowlim1 && x < lowlim_val1)
    x = lowlim_val1;
  if (use_highlim1 && x > highlim_val1)
    x = highlim_val1;
  diff = x - mean1;
  scale1 += diff * diff;
  x = val2[i];
  if (use_lowlim2 && x < lowlim_val2)
    x = lowlim_val2;
  if (use_highlim2 && x > highlim_val2)
    x = highlim_val2;
  diff = x - mean2;
  scale2 += diff * diff;
  }

scale1 = width * sqrt (scale1 / n_cases); // User param times standard deviation
scale2 = width * sqrt (scale2 / n_cases);

if (scale1 < 1.e-30) // Should never happen, but user may be careless
  scale1 = 1.e-30;

if (scale2 < 1.e-30)
  scale2 = 1.e-30;
```

We do an initialization that, in a sense, may not always be required. Code that allows a user to abort the later computation of grid (which can be slow for numerous cases and high resolution) is not shown here. However, most programmers will want to include an abort option to placate impatient users. Whatever fraction has been completed prior to interruption should be displayed. Thus, we initialize the entire display grid to zero in order to avoid nonsense numbers during display.

Also, we zero the total joint probability for scaling later. This is not used for display at all. However, the scaling described later is useful if the programmer wants to print some numeric values for the user.

```
for (i=0; i<res*res; i++)
  grid[i] = 0.0;   // Avoid nan in case user aborts

total_joint = 0.0;     // Used for printing numbers later, not display
```

The core computation is now performed. This computes the basic display grid, using Equations (3.1) through (3.4). Later, we'll do additional post-processing. But first, we handle the basics. Actually, we display the log of some quantities, which results in a much more interpretable image.

```
for (horz=0; horz<res; horz++) {                // Left to right across display
  x = xmin + horz * (xmax - xmin) / (res - 1);  // Map display horizontal to x value
  for (vert=0; vert<res; vert++) {              // Bottom to top of display
    y = ymin + vert * (ymax - ymin) / (res - 1); // Map display vertical to y value
    xmarg = ymarg = joint = 0.0;                // Will sum Equations 3.2 and 3.4
    for (i=0; i<n_cases; i++) {                 // Sum these two equations
      xdiff = (val1[i] - x) / scale1;           // d in Equations 3.1 and 3.3
      ydiff = (val2[i] - y) / scale2;
      xmarg += exp (-0.5 * xdiff * xdiff);      // Sum Equation 3.2
      ymarg += exp (-0.5 * ydiff * ydiff);
      joint += exp (-0.5 * (xdiff * xdiff + ydiff * ydiff));   // Sum Equation 3.4
    }
    xmarg /= n_cases * scale1 * root_two_pi;  // Complete Equation 3.2
    ymarg /= n_cases * scale2 * root_two_pi;
    joint /= n_cases * scale1 * scale2 * two_pi; // Complete Equation 3.4
```

```
if (xmarg < 1.e-50)                    // Do not allow zero denominator later
  xmarg = 1.e-50;
if (ymarg < 1.e-50)
  ymarg = 1.e-50;
if (joint < 1.e-100)
  joint = 1.e-100;

if (type == TYPE_DENSITY)
  grid[vert*res+horz] = log (joint);

else if (type == TYPE_MARGINAL)
  grid[vert*res+horz] = log (xmarg) + log (ymarg);

else { // INCONSISTENCY or MUTUAL INFORMATION
  numer = joint;
  if (numer < 1.e-100)
    numer = 1.e-100;
  denom = xmarg * ymarg;
  if (denom < 1.e-100)
    denom = 1.e-100;
  grid[vert*res+horz] = log (numer) - log (denom); // Eq (3.5) without abs value
                                                   // We'll do Abs Val later

  if (type == TYPE_MI) {           // If user wants mutual information
    total_joint += numer;          // Not used for display but useful for numbers
    grid[vert*res+horz] *= numer;  // This term in Equation (3.6)
    }
  } // Inconsistency or mutual information
  } // For vert
} // For horz
```

In the previous code, we actually compute the log of the density and marginal product when these quantities are to be displayed. I have found that this helps visual appeal. Feel free to experiment with displaying raw values or using other transformations.

The hard work is done. However, we perform some post-processing to improve the quality of the display as well as to optionally print a few numeric values that may be of interest to the user.

First, we handle displaying the contribution to mutual information. In the prior code block we computed the total joint probability. It's tempting to think this should sum to one, but remember that we are not summing across discrete categories; we are summing an approximate continuous density across a discrete grid, so the sum depends on the resolution. The following code divides the contributions to mutual information by this total as a form of normalization. This will not affect the display, but the sum of these normalized values, totalMI, is a specialized measure of mutual information that may be of interest to users for comparisons.

We also keep track of the point (maxMIx, maxMIy) in the domain at which the mutual information contribution is greatest, as well as the value (maxMI) of this maximum. I apply a special transformation to maxMI that accentuates sharply localized features. Recall (on page 19) that totalMI cannot be negative, and it will be zero only if the sample demonstrates perfect independence between the variables. In the extreme limiting case that all of the contribution comes from a single grid entry, unnormalized maxMI=totalMI. In this case, normalized maxMI=res*res.

```
if (type == TYPE_MI) {   // If user wants mutual information
   totalMI = 0.0;              // Not used for display, only optional printing
   maxMI = -1.e100;        // Ditto
   for (horz=0; horz<res; horz++) {
      x = xmin + horz * (xmax - xmin) / (res - 1);   // X value at this display position
      for (vert=0; vert<res; vert++) {
         y = ymin + vert * (ymax - ymin) / (res - 1); // And Y value
         grid[vert*res+horz] /= total_joint; // Normalize (does not impact display)
         totalMI += grid[vert*res+horz];    // Guaranteed non-negative
         if (grid[vert*res+horz] > maxMI) {
            maxMI = grid[vert*res+horz];
            maxMIx = x;
            maxMIy = y;
            }
         }
      }
   if (totalMI > 0.0)
      maxMI *= res * res / totalMI;
   else
      maxMI = 0.0;
   }
```

Now we consider displaying marginal inconsistency. The mutual information code in the prior section has no impact whatsoever on the display; it is strictly for producing some numerical values that may interest the user. This inconsistency code is the opposite; no numeric values for the user are computed, and the nature of the display itself is changed.

A significant problem with displaying raw values of the inconsistency given by Equation (3.5) on page 171 is that positive (concentration) and negative (sparsity) values are generally nonsymmetric. This has different implications depending on whether we take the absolute value shown in that equation and discussed in that section. For an effective visual display...

- If we do *not* take absolute value, we would like for inconsistency values of zero (the joint density equals the product of the marginals, indicating "normal" concentration) to have a visual appearance in the *center* of the display range.

- If we *do* take absolute values, we want "normal" regions displayed at one extreme and "abnormal" regions at the opposite extreme.

To satisfy these goals, we scale positive and negative values separately. Also, in this code we implement the absolute value shown in Equation (3.5) but not performed earlier when grid was computed. Some developers might find it more informative to refrain from taking the absolute value, for the reasons discussed earlier. I like it.

```
if (type == TYPE_INCONSISTENCY) {  // If user wants marginal inconsistency
  max_pos = max_neg = 1.e-20;
  for (i=0; i<res*res; i++) {
    if (grid[i] > 0.0 && grid[i] > max_pos)
      max_pos = grid[i];
    if (grid[i] < 0.0 && (-grid[i]) > max_neg)
      max_neg = -grid[i];
  }
  for (i=0; i<res*res; i++) {
    if (grid[i] > 0.0)
      grid[i] /= max_pos;
    if (grid[i] < 0.0)
      grid[i] /= -max_neg;       // Apply absolute value shown in Equation (3.5)
  }
}
```

A common technique for enhancing the visibility of differing tones or colors is *histogram equalization.* This technique applies a nonlinear transform to the data in such a way that every possible displayed tone or color occurs in the display in approximately equal quantity. The effect of this transformation is usually that small changes in the data are made more visible, while simultaneously reducing the prominence of large changes.

Recall that we allocated grid to be twice as long as needed. We'll now use the second half as scratch storage for sorting the grid values. The sorting routine qsortdsi() simultaneously moves the index keys, so after sorting we know the rank of each value. The result of this mapping code is that each entry in grid is from zero to one according to the fractile of the original value.

We apply one last optional transform. If the user requests that the boundary between large (anomalous) and not-so-large values be sharpened, we cube each entry. The result is that only values near the upper limit keep their vaunted position; lower values are pushed toward zero. This makes areas of unusually large concentration stand out from the background.

```
if (hist) {
  for (i=0; i<res*res; i++)
    keys[i] = i;

  sorted = grid + res * res; // Use last half for scratch
  memcpy (sorted, grid, res * res * sizeof(double));
  qsortdsi (0, res * res - 1, sorted, keys);

  for (i=0; i<res*res; i++)
    grid[keys[i]] = (double) i / (res * res - 1.0);

  if (sharpen) {
    for (i=0; i<res*res; i++)
      grid[i] = grid[i] * grid[i] * grid[i];
    }
} // Histogram equalization
```

If the user does not request histogram equalization, all we do is linearly rescale the values. This is more "authentic" in the sense that the display, whether in terms of tone or color, linearly reflects the grid values. The potentially extreme nonlinearity of histogram equalization can easily distort the visual perception of inconsistencies.

Note that the rescaling to 0-1 done here is not based on the extremes in grid. It is not unusual for there to be one or a few outliers, which would result in undue compression of the mapping. Rather, we discard the 1 percent largest and smallest values in grid and rescale so as to map those slightly narrower extremes to the display extremes of zero and one.

We also implement the optional sharpening discussed in conjunction with the prior code block.

```
else { // We scale by using ALMOST extremes
  sorted = grid + res * res; // Use last half for scratch
  for (i=0; i<res*res; i++)
    sorted[i] = grid[i];
  qsortd (0, res * res - 1, sorted);

  i = (int) (0.01 * res * res);
  smallest = sorted[i];               // Ignores smallest one percent
  largest = sorted[res*res-i-1];    // And largest
  mult = 1.0 / (largest - smallest + 1.e-20);     // Insure against largest=smallest

  for (i=0; i<res*res; i++) {
    grid[i] = mult * (grid[i] - smallest);
    if (grid[i] > 1.0)              // Happens for largest one percent
      grid[i] = 1.0;
    if (grid[i] < 0.0)              // Happens for smallest one percent
      grid[i] = 0.0;

    if (sharpen)
      grid[i] = grid[i] * grid[i] * grid[i];
  }
} // No histogram equalization
```

We're almost done. In most cases, the grid entries are now ready for display. However, users who want to highlight certain features, possibly for a demonstration or publication, may want to massage the display by shifting, compressing, or expanding the range of tones or colors. We provide the user with two parameters to accomplish this:

- ***Shift*** moves the overall display range. A positive value shifts the tones in the "high" direction, and negative shifts tones toward the "low" direction. The default of zero produces no change.

- ***Spread*** expands or compresses the range of the display. The default of
  zero produces no change. Negative values are legal but rarely useful, as
  this compresses variation into a narrow range, making discrimination
  difficult. Positive values, rarely beyond five or so, expand the center
  of the display range while squashing the extremes. This emphasizes
  features in the interior of the grid range, at the expense of the extremes.

Recall that grid ranges from zero to one. Close examination of the expansion section of
the following code shows that if spread is zero, no change in grid will occur. If grid[i]=0.5, it will
remain unchanged, regardless of spread. As grid[i] moves away from 0.5, its transformed value
will do the same monotonically, with the rate determined by the multiplier.

```
if (spread >= 0.0)           // Usual situation
   mult = spread + 1.0;
else                         // Rarely useful, as it generally degrades the display
   mult = 1.0 / (1.0 - spread);

for (i=0; i<res*res; i++) {
   grid[i] += 0.01 * shift;   // This is where the display is shifted; 0.01 is arbitrary
   if (grid[i] < 1.e-12)      // Needed for log below
      grid[i] = 1.e-12;
   if (grid[i] > 1.0 - 1.e-12)  // Ditto
      grid[i] = 1.0 - 1.e-12;

   if (grid[i] <= 0.5)
      grid[i] = 0.5 * exp (mult * log (2.0 * grid[i]));
   else
      grid[i] = 1.0 - 0.5 * exp (mult * log (2.0 * (1.0 - grid[i])));
}
```

Comments on Showing the Display

I don't present any code for displaying grid. This is because display code is highly
implementation-specific. My own code in the DATAMINE program uses numerous
Windows API calls that might be unacceptable to other programmers. I choose to do
this because it allows me to easily place scales and text on the display, at the expense of

taking a relatively long time to display, as it's done one pixel at a time. Nevertheless, here are a few issues to keep in mind when writing your own code to display grid:

- Grayscale is good for publication in black-and-white formats, but colors are more visually pleasing. Avoid red-versus-green, as this is the most common form of color blindness. Red-versus-blue is good, as is yellow (red+green) versus blue. You can compute levels as follows:

  ```
  red_level = (int) (val * 255.99);
  blue_level = (int) ((1.0-val) * 255.99);
  SetPixel (..., RGB (red_level, red_level, blue_level));
  ```

- Computing grid at full display resolution is impractical. Linearly interpolate in both directions. Bivariate linear interpolation algorithms are readily available and not shown here, as the exact implementation depends on the display method. Windows provides a routine (StretchDIBits) that rapidly does the interpolation, but labeling the display becomes much more difficult.

- When printing the display (as opposed to displaying it on a monitor), be aware that many printers have extremely high resolution, making interpolation much too slow. In this case, print small rectangles instead of individual pixels.

CHAPTER 4

Fun with Eigenvectors

Suppose we measure the height and weight of a collection of people. We could make a plot of the results, using an asterisk for each person. The horizontal position is determined by the person's height, and the vertical position is determined by the person's weight. The resulting plot might look something like that shown in Figure 4-1.

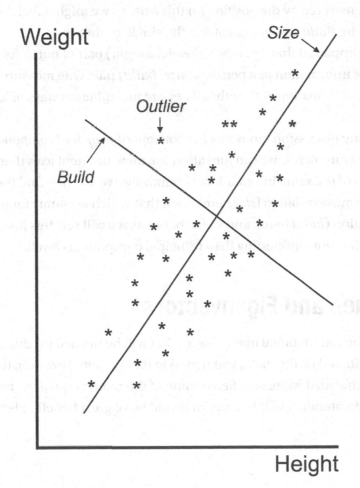

Figure 4-1. *Simple principal components*

© Timothy Masters 2018
T. Masters, *Data Mining Algorithms in C++*, https://doi.org/10.1007/978-1-4842-3315-3_4

Not surprisingly, these two measurements are highly correlated; tall people tend to weigh more than short people. Of course, the correlation is not perfect; some people are built differently from others.

One thing that jumps out of a plot of highly correlated variables is that there exists a *principal axis*, the direction in which most variation lies. In this example, the principal axis can be labeled the *size* of the person. For each of these people, we can drop a line perpendicular to the *size* axis and see where this line intersects the axis. The location of this point, measured along this axis, is a good measurement of the "size" of the person.

But there is another dimension to consider. A parsimonious way to measure this other dimension is to consider the axis perpendicular to the first. In this example, this second axis depicts discrepancies between a person's actual weight and the weight expected from their height. Is a person unusually heavy or light for their height? This is the question answered by the position on this axis, so we might label this axis *Build*. Notice that it is the *Build* axis that identifies the single outlier.

It should be apparent that a person's (*height, weight*) pair of numbers provides exactly the same information as a person's (*size, build*) pair. One measurement pair is a simple linear transformation of the other. They are just different ways of looking at the same information.

The preceding discussion motivates the concept of *principal components*. Given multivariate measurements, we can find alternate measurement axes that capture different aspects of the same information. Commonly, we will first find the axis that accounts for the most variation (*size* here), then that which accounts for most of the remaining variation (*build* here), and so forth. But as we will see, this just scratches the surface. Things far more interesting than principal components await.

Eigenvalues and Eigenvectors

We begin with the foundational mathematics that will be needed for this chapter. If you are totally intimidated by the math, you may skip this section. However, this math is not particularly advanced, despite how fierce some of the matrix equations may look, and at least a basic understanding of this material would be of great benefit. Please try.

Suppose $\mathbf{A}$ is a p by p matrix, $\mathbf{x}$ is a column vector p long, and λ is a scalar. Then $\mathbf{x}$ is said to be an *eigenvector* of $\mathbf{A}$, and λ its associated eigenvalue, if and only if Equation (4.1) holds.

$$\mathbf{Ax} = \lambda\mathbf{x} \tag{4.1}$$

It should be apparent that any multiple of $\mathbf{x}$ is also an eigenvector; the concept of eigenvector applies only to direction, not length. Therefore, a common convention when computing eigenvectors is to normalize them to unit length. We will do so, and always make this assumption.

Although not critical to the topic at hand, it is interesting to note a simple geometric interpretation of eigenvectors. Multiplication of a vector by a matrix will, in general, rotate the vector. But the eigenvectors of a matrix have the property that when multiplied by the matrix, they do not change direction. They are a sort of "stationary" direction for the matrix.

The relevance of eigenvectors to this chapter's material comes from another of their properties. Suppose we observe $\mathbf{x}$, a p-vector drawn from a standardized multivariate normal distribution. In other words, each of its components has a normal distribution with mean zero and unit variance. The covariance matrix is also (due to the standardization) the correlation matrix. Call it $\mathbf{R}$. Let $\mathbf{V}$ be a p by m matrix, with $m<=p$. Consider the new random vector, m long, defined by Equation (4.2).

$$\mathbf{y} = \mathbf{V}'\mathbf{x} \tag{4.2}$$

It can be shown (though we will not do so here, as the derivation is widely available elsewhere) that the covariance matrix of $\mathbf{y}$ is given by Equation (4.3).

$$\mathbf{C} = \mathbf{V}'\mathbf{RV} \tag{4.3}$$

Let's explore some desirable properties of $\mathbf{V}$, properties that will provide useful properties of $\mathbf{y}$. Suppose for the moment that $m=1$; $\mathbf{V}$ has just a single column. Then the "covariance matrix" $\mathbf{C}$ is a single number, the variance of $\mathbf{y}$. A set of weights for the members of $\mathbf{x}$ that results in $\mathbf{y}$ having the maximum possible variance has great intuitive appeal because this is the transformation that, in a sense, captures the most information about variation in $\mathbf{x}$. See Figure 4-1 on page 185 and consider the *size* dimension.

Obviously, multiplying the weights by a constant will multiply the variance of **y** by the square of that constant, so we must impose some sort of normalization on **V**. The most sensible restriction is that the square of the components of **V** sum to one. Equivalently, the length of the column is one.

It turns out that this single column of **V** is the eigenvector of **R** that corresponds to the largest eigenvalue. The proof of this fact is not difficult, but because it is tedious and easily available elsewhere, we dispense with its presentation.

Now suppose that we let $m=2$, so **V** has two columns. We let the first column be the eigenvector corresponding to the largest eigenvalue, as just described. How can we define the second column so that the second component of **y** is orthogonal to the first component (the two components of **y** are independent) and this second component of **y** has the maximum possible remaining variance? Not surprisingly, this second column is the eigenvector of **R**, which corresponds to the second-largest eigenvalue. This pattern repeats for all p possible columns of **V**. Thus, the eigenvectors of **R** provide the transformation matrix for mapping the standardized, likely correlated **x** variables to new independent **y** variables with the property that they capture the most, second most, and so forth, variance in **x**.

Principal Components (If You Really Must)

Many developers take advantage of these orthogonal and descending variance properties to compute and employ the *principal components* of a dataset. They may have a collection of variables so large as to be unwieldy. By finding the eigenvalues and vectors of the correlation matrix, the developer can compute a much smaller set of new variables that capture the majority of the variation in the original set. For example, one might begin with 100 variables. The first principal component may account for perhaps 20 percent of their total variance, the second another 10 percent, and so forth. It may turn out that just 15 new variables can capture as much as 90 percent of the original set's variance. This would not be terribly unusual, and it is enticing.

Beware of that enticement. There is one important caveat about using principal components to whittle down the number of variables in an application: we likely don't know in advance which components (if any!) convey the information in which we are interested. It is the case that in many applications, early components convey most of the useful information, while noise tends to be concentrated in the late principal components. But this is far from universal. For example, turn back to page 185 and look at Figure 4-1. Suppose our goal is to predict how well a person would do in a football game. Clearly, the *size* dimension would be far more valuable than the *build* dimension. But the opposite would be true if we were trying to predict likelihood of developing diabetes. So, the very real danger of variable reduction via principal components is that we may discard the dimensions that are most important to our application!

If you do choose to be brave and compute the principal components of your standardized variables by weighting them according to the eigenvectors, you would generally do well to take one more step. The variance of each computed principal component is the eigenvalue associated with that eigenvector. Thus, before doing the weighting (Equation 4.2), it makes sense to divide each eigenvector by the square root of its eigenvalue. By doing so, the variance of each component is standardized to one. This equalization of variation is appreciated by most data mining and model training algorithms.

The Factor Structure Is More Interesting

The world is filled with textbooks (mostly in the field of psychology) that explore in detail methods for using principal components and factor models (page 221) to discover and label dimensions of interest. These techniques can be useful, and I certainly will not scorn them. But such labeling techniques are not among my main reasons for computing eigenvalues and vectors of a dataset and will receive only passing note in the next section. If you desire a more complete discussion, you are encouraged to explore this material elsewhere. "Modern Factor Analysis" by Harry Harmon, though not so modern any more, is an exceptionally thorough and well written reference for the core material.

What particularly interests me in regard to eigenstructure as related to data mining is how each of our (potentially numerous) measured variables relates to the dominant axes of variation, *whatever these axes may represent*. Of course, finding descriptive names for axes of variation can often be interesting and useful; we'll briefly explore a contrived example in the next section. But what is usually of greatest importance is the *correlation* between each variable and each principal component (or at least those corresponding to the largest eigenvalues). The axes may possibly be unnamed or even unnameable by mere mortals; psychologists love giving them names, while I, as a data miner, don't usually care as much. But once again, I emphasize that I do not disparage a quest for names; we'll see an example in the next section in which naming can be interesting. It's just that one should never be discouraged if a descriptive name does not pop out of the data; names are usually of secondary importance to data miners.

The matrix of variable/component correlations is called the *factor structure matrix* and is computed by multiplying each normalized (unit length) eigenvector by the square root of its corresponding eigenvalue. (For historical and theoretical reasons best omitted here, this matrix is also called the *factor loading* matrix.) Now let's explore a simple, contrived example of how the factor structure can reveal interesting relationships between variables.

A Simple Example

Using many years of a common equity market index, I computed a set of ten trend measurements as well as a set of ten corresponding volatility measurements with a moving window. In other words, for a 50-day window I looked at the first 50 days in the price history and computed a numeric measurement of the trend within that window. I also computed a measure of price volatility within that same window. Then I advanced the window forward in time by one day and did the same. These trend and volatility measurements were done with window lengths of 50, 51, 52, ..., 59 days, giving a total of ten different window lengths. Obviously, there will be huge correlation between variables for these different window sizes, because the lengths are so similar. This was deliberate on my part so as to produce a clear demonstration of the technique.

The table shown next lists the four largest eigenvalues, along with their corresponding factor structures. The *Cumulative* row shows the cumulative percent of variation captured by each column and is computed as the cumulative sum of eigenvalues divided by the total of all eigenvalues.

Eigenvalue	12.939	6.900	0.090	0.052
Cumulative	64.693	99.193	99.643	99.904
TREND_50	0.7829	0.6040	0.1416	0.0356
TREND_51	0.7893	0.6030	0.1115	0.0280
TREND_52	0.7949	0.6010	0.0796	0.0201
TREND_53	0.7999	0.5980	0.0466	0.0119
TREND_54	0.8041	0.5939	0.0133	0.0035
TREND_55	0.8076	0.5890	-0.0195	-0.0052

TREND_56	0.8105	0.5831	-0.0510	-0.0140
TREND_57	0.8127	0.5765	-0.0805	-0.0229
TREND_58	0.8144	0.5692	-0.1075	-0.0319
TREND_59	0.8155	0.5613	-0.1319	-0.0409
VOL_50	-0.8214	0.5570	0.0461	-0.1036
VOL_51	-0.8188	0.5652	0.0385	-0.0863
VOL_52	-0.8160	0.5727	0.0287	-0.0644
VOL_53	-0.8127	0.5796	0.0172	-0.0391
VOL_54	-0.8090	0.5861	0.0052	-0.0124
VOL_55	-0.8047	0.5919	-0.0072	0.0140
VOL_56	-0.8003	0.5969	-0.0198	0.0393
VOL_57	-0.7954	0.6012	-0.0316	0.0626
VOL_58	-0.7902	0.6051	-0.0415	0.0826
VOL_59	-0.7845	0.6086	-0.0496	0.0983

Now let's explore some properties of this table. Recall that these are correlations. For example, the variable TREND_51 has a correlation of 0.1115 with the third principal component. Here are some notable features of this table:

- The first principal component, *a single new variable*, captures almost two-thirds (64.693 percent) of the entire variation inherent in the complete set of 20 variables.

- If we throw in the second principal component, we've garnered more than 99 percent of the variation.

- The dominant component, which accounts for almost two-thirds of the total variation of all variables across the dataset, is fascinating, as it is a contrast between trend and volatility. Large values of this principal component correspond to conditions within the window of strong upward trend (correlation with trend is about 0.8) combined with low volatility (correlation with volatility of about -0.8). Conversely, unusually small values of this first principal component correspond to strong downward trend and high volatility. So we might think of this new variable as telling us whether the market is engaged in a peaceful rise versus a turbulent plunge.

- The second component indicates the degree and direction of departures from the dominant behavior embodied in the first component, as it is moderately positively correlated with all variables. Large values of this second principal component identify times when the market is trending upward but with high volatility. Similarly, very negative values signify a falling market with low volatility.

- The third, very minor, principal component distinguishes between effects that are happening for short versus long windows, with one type of interaction between trend and volatility.

- The fourth also distinguishes between short versus long, but with the opposite trend/volatility relationship. By now we've left less than one-tenth of 1 percent of the total 20-variable variation on the table!

Rotation Can Make Naming Easier

I know I keep stating that naming axes is of secondary importance, and I hesitate to dwell on the topic too much. But there is one issue that should be at least mentioned, lest I be accused of negligence.

We saw in the prior section that just the two most dominant principal components account for more than 99 percent of the total variation in all 20 variables. And in this contrived example, the meanings of these two components were obvious. But this was the case only because I deliberately employed two sets of variables that enjoyed high within-set correlation. Usually we are not so fortunate, and we will encounter factor structure members (correlations) along a continuum. This can make naming, or at least guessing properties of the components, difficult. There is a technique called *varimax rotation* (other, less popular methods also exist) that can make interpretation easier. With no loss of information, this algorithm rotates the axes in such a way that correlations are driven to extreme values: +/- 1 and 0. By reducing the number of intermediate correlations, interpretability is often enhanced. The following table shows the first two principal components after varimax rotation:

	Commun	Pct	
TREND_50	97.78	0.1277	0.9805
TREND_51	98.66	0.1329	0.9844
TREND_52	99.31	0.1383	0.9869
TREND_53	99.73	0.1439	0.9882

TREND_54	99.93	0.1498	0.9884
TREND_55	99.91	0.1558	0.9873
TREND_56	99.69	0.1619	0.9852
TREND_57	99.29	0.1682	0.9821
TREND_58	98.72	0.1745	0.9781
TREND_59	98.01	0.1809	0.9733
VOL_50	98.48	-0.9748	-0.1858
VOL_51	98.99	-0.9789	-0.1782
VOL_52	99.38	-0.9822	-0.1709
VOL_53	99.65	-0.9847	-0.1637
VOL_54	99.79	-0.9866	-0.1565
VOL_55	99.79	-0.9877	-0.1493
VOL_56	99.67	-0.9881	-0.1427
VOL_57	99.41	-0.9877	-0.1362
VOL_58	99.05	-0.9867	-0.1298
VOL_59	98.59	-0.9853	-0.1232

We have three columns. Look at the last two columns. These correspond to the first two principal components, after rotation. Note that one column assigns large magnitude weights to the trend variables and small weights to the volatility. The other column does the opposite. This has a benefit and a cost. The benefit is that naming these two axes is suddenly a lot easier: one column can clearly be named *Trend* and the other named *Volatility*. But the cost is that we have lost the ordering property. We can no longer say that one of these components is dominant, and so forth.

The first column in this table is especially important. When we discard principal components (in this case, we discarded 18 of the 20, keeping only the first two for rotation), we inevitably lose some of the information in the original variables. The *communality* of a variable, usually expressed in percent, is the fraction of the variance of that variable that is encapsulated in the components that are kept. It is computed by summing the squares of the factor correlations across that variable's row. For example, in this case we see that the first two principal components contain 97.78 percent of the variance of the TREND_50 variable, and this is 0.1277 squared plus 0.9805 squared. *Knowing the communalities can help us identify variables that are under-represented in the principal components that we kept.*

This discussion of factor structure interpretation, and especially rotation, has been perhaps shamefully brief. If you are rolling your eyes in bafflement right now, I express a somewhat hesitant apology. However, this was a deliberate choice. The general topic of identifying axes by name or property is not a major activity in my own data mining experience, and hence it is not a major topic in this chapter. Moreover, these topics are covered in excruciating detail in numerous other texts, so expounding on them in detail would be a waste of valuable trees. At least this limited presentation provides an overview of what can be done, so that interested readers can look elsewhere for more details. We will soon see much more important (in my opinion!) uses for eigenvectors.

Code for Eigenvectors and Rotation

Three files relevant to the prior discussion can be downloaded from my web site. These are the following:

- *EVEC_RS.CPP*: This is a ready-to-use C++ subroutine that computes eigenvalues and (optionally) eigenvectors of a real symmetric matrix.

- *AN_EIGEN.TXT*: This is essential code fragments that fetch data from a database and compute the factor structure information.

- *AN_ROTATE.TXT* : This is essential code fragments that perform varimax rotation of a factor structure.

None of these routines will be examined in full detail in this text because the algorithms are standard and widely available elsewhere; there is no point in being redundant. But each will be presented in sufficient detail so you can understand how to use them in your own code.

Eigenvectors of a Real Symmetric Matrix

This subroutine, EVEC_RS.CPP, should be ready to compile with any C++ compiler. It uses a reliable and efficient standard algorithm for eigenvalue and optional eigenvector computation for a real symmetric matrix. First, the matrix is transformed to tridiagonal form using the Householder method. Then the eigenvalues are computed using the QL algorithm with implicit shifts. If eigenvectors are also desired, the rotations are cumulated. This cumulation is an expensive process, so eigenvectors should be computed only if they are needed.

Note that several theoretically superior methods (*divide-and-conquer, MRRR*) are now available. However, they are still n-cubed operations and differ in speed only by a modest factor. They are tremendously more complex than the method given here, and simple, thoroughly vetted and documented C++ source code for them is difficult to obtain. FORTRAN versions are available in LAPACK.

This routine is called as follows:

```
int evec_rs (double *mat_in, int n, int find_vec,
             double *vect, double *eval, double *workv)
```

- *mat_in*: Square input matrix, with columns changing fastest. The upper-right triangle (column greater than row) is ignored and may contain any values. This input matrix is left unchanged. If you want to modify the source code for more compact storage ((1,1), (2,1), (2,2), ...), you should find it easy to do so, as this input matrix is simply copied into working storage and thereafter ignored.

- *n*: Size of the matrix.

- *find_vec*: If nonzero, the eigenvectors will also be computed. This tremendously increases compute time.

- *vect*: Square matrix n by n. The eigenvectors are output here if *find_vec* is nonzero. Even if *find_vec* is zero, this matrix must still be supplied, because it is used for scratch storage. It is legal to use the same matrix for *mat_in* and *vect*, in which case the input matrix is replaced.

- *eval*: Output of eigenvalues, sorted descending

- *workv*: Scratch vector n long

This routine returns the number of eigenvalues that, due to convergence problems, were not able to be computed. I've tested it with thousands of matrices, up to 5000 by 5000, many very ill conditioned, and I've never seen it fail; in my experience, it *always* returns zero, indicating success. However, there is the theoretical possibility of failure, so I account for this possibility in my code.

Factor Structure of a Dataset

The file AN_EIGEN.TXT contains code fragments that illustrate the essential aspects of computing the factor structure of a dataset. The following variables appear in this code:

- *n_cases*: Number of cases (rows) in database

- *n_vars*: Number of columns in database (not all of which may take part)

- *database*: All data is here, an *n_cases* by *n_vars* matrix

- *npred*: Number of predictors (variables) taking part in this analysis

- *preds[]*: Array *npred* long that identifies the columns in the database for the variables to be used in this analysis

The first step is to allocate memory. The two variables that begin *eigen_* are global because further user operations may be performed on them. The other allocations are temporary for this routine.

```
cumulative = (double *) MALLOC (npred * sizeof(double));
covar = (double *) MALLOC (npred * npred * sizeof(double));
evals = (double *) MALLOC (npred * sizeof(double));
structure = (double *) MALLOC (npred * npred * sizeof(double));
means = (double *) MALLOC (npred * sizeof(double));
stddev = (double *) MALLOC (npred * sizeof(double));
```

Compute the means and standard deviations so we can standardize the data. Note how we extract the required data from the database.

```
for (i=0; i<npred; i++)
  means[i] = stddev[i] = 1.e-60; // Must not divide by zero later

for (i=0; i<n_cases; i++) {
  for (j=0; j<npred; j++)
    means[j] += database[i*n_vars+preds[j]];
  }

for (j=0; j<npred; j++)
  means[j] /= n_cases;
```

```
for (i=0; i<n_cases; i++) {
  for (j=0; j<npred; j++) {
    diff = database[i*n_vars+preds[j]] - means[j];
    stddev[j] += diff * diff;
    }
  }

for (j=0; j<npred; j++)
  stddev[j] = sqrt (stddev[j] / n_cases);
```

Compute the covariance matrix, which is also a correlation matrix because the variables have been standardized. We do not have to compute the upper triangle because the matrix is symmetric, nor do we compute the diagonal, because it is identically 1.0 due to standardization. Copying the triangle at the end is needed only if required by a different eigen routine.

```
for (i=1; i<npred; i++) {
  for (j=0; j<i; j++)
    covar[i*npred+j] = 0.0;
  }

for (i=0; i<n_cases; i++) {
  for (j=0; j<npred; j++) {
    diff = (database[i*n_vars+preds[j]] - means[j]) / stddev[j];
    for (k=0; k<j; k++) {
      diff2 = (database[i*n_vars+preds[k]] - means[k]) / stddev[k];
      covar[j*npred+k] += diff * diff2;
      }
    }
  }

for (j=0; j<npred; j++) {
  for (k=0; k<j; k++)
    covar[j*npred+k] /= n_cases;
  }
```

```
for (j=0; j<npred; j++) {
  covar[j*npred+j] = 1.0;                    // Definition, so not computed
  for (k=j+1; k<npred; k++)                  // Copying the other triangle is not needed
    covar[j*npred+k] = covar[k*npred+j];     // for evec_rs() and may be omitted
}
```

Compute the eigenvalues and vectors using our evec_rs() routine. In the previous code, we copied the computed lower-left triangle to the upper right. But our evec_rs() ignores that upper triangle, so those two lines of copying code may be omitted. They are shown here only because some other routines may require the entire matrix. Then we compute the cumulative eigenvalues and divide by the sum to express the cumulative values as percents. It may rarely happen that tiny floating-point errors result in slightly negative eigenvalues, a theoretical impossibility here, so we enforce non-negativity.

```
evec_rs (covar, npred, 1, structure, evals, means);

sum = 0.0;
for (i=0; i<npred; i++) { // We display cumulative eigenvalues
  if (evals[i] < 0.0) // Happens only from tiny fpt errors
    evals[i] = 0.0;
  sum += evals[i];
  cumulative[i] = sum;
}

for (i=0; i<npred; i++)   // Make it percent
  cumulative[i] = 100.0 * cumulative[i] / sum;
```

The last step is to multiply each eigenvector by the square root of its eigenvalue in order to get the factor structure (also called the *factor loadings* in some contexts). It may rarely happen that tiny floating-point calculations result in correlations trivially beyond +/-1. To prevent this nonsense, we enforce theory.

```
for (i=0; i<eigen_npred; i++) {
  for (j=0; j<eigen_npred; j++) {
    structure[i*npred+j] *= sqrt(evals[j]);
    if (structure[i*npred+j] < -1.0) // In a perfect fpt world this would never happen
      structure[i*npred+j] = -1.0;
```

```
    if (structure[i*npred+j] > 1.0)
      structure[i*npred+j] = 1.0;
    }
  }
```

Varimax Rotation

The varimax rotation algorithm is iterative, but it converges quickly in nearly all cases. It sweeps through every pair of columns (correlations of a factor with all variables) and explicitly computes the angle of rotation that maximizes a measure of optimality, where optimality is (loosely) defined as the correlations being as near +/-1 and 0 as possible. Of course, each time this pairwise rotation is done, optimality of a prior pair is impaired. Thus, multiple sweeps must be done until an entire set of all pairs has negligible change.

The exact equations for computing the optimal rotation angle are fierce and widely available in other references, so they will not be reproduced here. However, we will work through the code provided in AN_ROTATE.TXT so that you understand how to use this code in your own project. In this code, n_kept is the number of dominant (earliest) factors that we will rotate. It must be at least two and at most npred.

The first step is to compute the square root of the communalities. Recall (page 193) that the communality of a variable is the fraction of that variable's variance that is accounted for by the factors that are retained. After computing these, we temporarily scale the factor structure. When rotation is complete, we will reverse this scaling to restore the correct communalities; rotation does not change communality. The original version of varimax rotation did not perform this scaling, but much experience indicates that it improves interpretability.

```
for (i=0; i<npred; i++) {
  sum = 0.0;
  for (j=0; j<n_kept; j++)
    sum += structure[i*npred+j] * structure[i*npred+j];
  comm[i] = sqrt (sum);
  }
```

```
for (i=0; i<npred; i++) {
  sum = comm[i];
  for (j=0; j<n_kept; j++)
    structure[i*npred+j] /= sum;
}
```

Now we have the main outer loop that repeatedly sweeps through all pairs of columns (factors) until a complete sweep results in no change. We impose an iteration limit of 100 as cheap insurance against an endless loop. In practice, we never come even close to this limit. We set the convergence flag to *True* (1) before we start the pairwise sweeping. If even a single rotation is done during a sweep, this flag is reset to *False*. At the end of the outer iteration loop, if the flag is still *True*, we break out of the loop.

```
for (iter=0; iter<100; iter++) { // limit is for safety and should never come even close
  converged = 1;          // We'll reset this if an adjustment is made
  for (first_col=0; first_col<n_kept-1; first_col++) { // Do all pairs of cols
    for (second_col=first_col+1; second_col<n_kept; second_col++) {
      A = B = C = D = 0.0;   // We will sum these down the row (all vars)
```

At this point we have a pair of columns (first_col and second_col) that will be rotated. Now we have to figure out how much to rotate. Without delving into details that are tedious and widely available elsewhere, the idea is that there is an optimality criterion that we want to maximize. The derivative of this criterion with respect to the rotation angle phi will be zero at the maximum, and the second derivative will be negative. The angle that satisfies these two rules can be explicitly computed. To do so, sum down rows the quantities we will need to compute the rotation angle.

```
for (ivar=0; ivar<npred; ivar++) {      // Sum down all rows
  row_ptr = structure + ivar * npred;  // This var's row in structure matrix
  load1 = row_ptr[first_col];
  load2 = row_ptr[second_col];
  Uterm = load1 * load1 - load2 * load2;
  Vterm = 2.0 * load1 * load2;
  A += Uterm;
  B += Vterm;
  C += Uterm * Uterm - Vterm * Vterm;
  D += 2.0 * Uterm * Vterm;
  } // For ivar
```

```
numer = D - 2.0 * A * B / npred;
denom = C - (A * A - B * B) / npred;
phi = 0.25 * atan2 (numer, denom);   // This is the rotation angle
```

If the angle by which we are to rotate this pair of columns is tiny, there is no point bothering. Otherwise, do the rotation and reset the convergence flag to *False*.

```
if (fabs(phi) < 1.e-10)   // No point rotating this pair of columns if angle is tiny
   continue;              // So go on to the next pair of columns

sin_phi = sin (phi);
cos_phi = cos (phi);

for (ivar=0; ivar<npred; ivar++) {    // Rotate this pair of columns
   row_ptr = structure + ivar * npred; // This var's row in structure matrix
   load1 = row_ptr[first_col];
   load2 = row_ptr[second_col];
   row_ptr[first_col] =  cos_phi * load1 + sin_phi * load2;
   row_ptr[second_col] = -sin_phi * load1 + cos_phi * load2;
   }

converged = 0; // We just made an adjustment, so we are not converged

   } // For second column
  } // For first column

if (converged)
   break;
} // For iter (main outer loop)
```

The final step is to undo the communality scaling that we did at the start of this routine.

```
for (i=0; i<npred; i++) {
sum = comm[i];
for (j=0; j<n_kept; j++)
   structure[i*npred+j] *= sum;
}
```

Horn's Algorithm for Determining Dimensionality

Whether one wants to compute principal components or name axes, discover axes of variation without naming them, or employ the variable clustering technique described in the next section, it is important to be able to decide how many dimensions of the data are relevant. On page 190 we saw a simple contrived example in which twenty variables could be reduced to just two while retaining nearly all variation inherent in the set. For other datasets, it may be that little or no dimension reduction is possible. It would be nice to have a theoretically supportable method for determining the number of dimensions inherent in the data, with the assumption that discarded dimensions are just noise, devoid of useful information.

Of course, before pursuing this line of thought, we must once more emphasize that this is a potentially dangerous operation. We already saw in the height/weight example that opened this chapter, the *Size* dimension would likely be useful for assessing football performance, while the *Build* dimension would be applicable to diabetes screening. And in the example on page 190, it is clear that components past the strongly dominant first two also contain clearly identifiable information. So, dimension reduction is always fraught with the danger of discarding precisely the information most valuable to your application. With that caveat, we continue.

The traditional way to determine the appropriate number of dimensions is to plot the eigenvalues, left to right on the plot, in descending order. This is called a *scree plot*. Typically, the eigenvalues will drop off quickly at first and then form a knee and flatten. The developer visually determines the location of the knee and sets a cutoff at that number of components to retain. The problem with this approach is that it is inherently subject to human interpretation and bias.

A fairly justifiable approach, commonly used, relies on the fact that if the variables are completely independent (no dominant axes due to underlying components that impact multiple variables), then their theoretical correlation matrix will be an identity matrix, and hence all eigenvalues will equal 1.0. The degree to which the eigenvalues separate above and below 1.0 indicates the degree to which the measured variables are being driven by underlying common factors. This inspires a rule that says we should keep all principal components whose eigenvalues exceed 1.0.

The small but troubling problem with this rule is that for finite datasets, random variation will cause significant spreading of the eigenvalues, even if the data has been drawn from a population of truly independent variables. A better approach, especially if the number of cases is not enormous compared to the number of variables, is to use a Monte Carlo procedure to estimate the actual distribution of the ordered eigenvalues under the hypothesis that all variables are independent.

The paper [Horn, J. (1965). "A rationale and test for the number of factors in factor analysis." *Psychometrika*, 30(2), 179–185.] suggested that a large number (hundreds at least) of data matrices of the same size as that under study be generated, each being sampled from a population of independent variables. For each sample, compute and sort the eigenvalues of the correlation matrix. Then compute the average, across all samples, of each ordered eigenvalue. We would almost surely find that the average of the largest eigenvalue significantly exceeds 1.0, with subsequent ordered values similarly departing from theory. Then we use these averages as the cutoff thresholds, instead of the theoretical value of 1.0.

The actual algorithm is slightly different from what might be implied by the description just given. The issue is that random variation in the Monte Carlo procedure could result in gaps in the selection procedure. For example, if the ordered thresholds were directly applied, we might find that factors 1, 2, 3, and 5 are kept, with factor 4 falling under its threshold and hence rejected. So what is done is to use the thresholds as a stopping criterion: start at the largest eigenvalue and work downward, stopping the first time a threshold is violated.

Recent experience indicates that limiting users to the mean across Monte Carlo replications is overly restrictive. A more general approach is to let the user specify in advance a percentile across replications. For each ordered position, the specified percentile of that ordered eigenvalue is used as the threshold for rejection.

Code for the Modified Horn Algorithm

The stopping algorithm just described is simple to implement. Assume for the moment that we have used a Monte Carlo algorithm to compute the eigenvalue thresholds, and they are in the array thresh. So, thresh[0] contains the threshold for the largest eigenvalue, thresh[1] the threshold for the second-largest, and so forth. In the original Horn algorithm, thresh[0] would be the mean across all Monte Carlo replications of the largest eigenvalue, and so forth. In the more modern method that will be presented later, these would be a user-specified percentile of each ordered eigenvalue. To determine how many factors to retain, we can use the following trivial code:

```
for (n_kept=0; n_kept<npred; n_kept++) {
  if (evals[n_kept] <= thresh[n_kept])
    break;
  }
```

The trickier part is computing these thresholds. Conceptually, it's not difficult. But because we will be building correlation matrices and finding eigenvalues many times (typically several hundred or so), it behooves us to use multithreading so as to take advantage of modern multicore CPUs. This is the code that will now be presented. If you want to keep it simple and use a single thread should find it easy to do so.

Recall that Windows allows passing only a single parameter to a threaded routine, so we'd better make it a good one. In this case we will pass a pointer to a structure that contains everything needed. Here is this structure:

```
typedef struct {
  int nc;          // Number of cases
  int nv;          // Number of variables
  double *covar;   // Scratch for covariance matrix
  double *evals;   // Computed eigenvalues
  double *workv;   // Scratch vector for evec_rs()
  int ieval;       // Needed for placing result in all_evals
} MC_EVALS_PARAMS;
```

This is the routine that performs a single Monte Carlo replication. Single-threaded implementations will call it as many times as desired in a simple loop. Multithreaded applications such as the one presented here will run multiple copies simultaneously.

The first step is to fetch the items passed in the structure. This is for clarity only; a programmer could just as well directly reference the structure each time. I like my approach better. Note that we assign the evals and workv members to two different variables. Again, this is just for clarity. We will use these two vectors for different things at different times, so using context-appropriate names helps reduce confusion.

```
static unsigned int_stdcall evals_threaded (LPVOID dp)
{
   int i, j, icase, n_cases, n_vars;
   double *xvec, *means, *covar, xtemp, *evals, *workv;

   n_cases = ((MC_EVALS_PARAMS *) dp)->nc;
   n_vars = ((MC_EVALS_PARAMS *) dp)->nv;
   covar = ((MC_EVALS_PARAMS *) dp)->covar;
   xvec = evals = ((MC_EVALS_PARAMS *) dp)->evals;  // Borrow for computing covar
   sums = workv = ((MC_EVALS_PARAMS *) dp)->workv;  // Ditto
```

We will compute the lower-left triangle of the covariance (and then correlation) matrix of a standardized, uncorrelated normal random variable. The upper-right triangle is ignored by the evec_rs() routine that computes eigenvalues. So, begin by zeroing the areas where the mean and covariance will be cumulated.

```
   for (i=0; i<n_vars; i++) {
      sums[i] = 0.0;
      for (j=0; j<=i; j++)
         covar[i*n_vars+j] = 0.0;
      }
```

This loop generates the required number of cases. This should be the same as the number of cases in the dataset being analyzed. The function normal_pair() computes two standard (mean zero, unit variance) random numbers at a time, which is the most efficient way to do it. This function is provided in the file RANDOM.CPP, which is available for free download from my web site. The first loop within the icase loop constructs the random vector xvec.

```
   for (icase=0; icase<n_cases; icase++) {

      // Generate the random vector
      for (i=0; i<n_vars; i++) {
         if (i % 2 == 0)
            normal_pair (&xvec[i], &xtemp);
         else
            xvec[i] = xtemp;
         }
```

The second loop inside the icase loop cumulates the means and sum of squares. In a more general setting, we would want to make two passes through the data. The first pass cumulates the mean, and the second pass cumulates the sum of squared deviations from the mean. But that method, though most accurate, requires storing the entire dataset. As it may be huge, and we would need a separate dataset for each of the multiple threads, it would be nice to avoid this storage. It happens that in this application, we can get away with the otherwise dangerous "no-store" method. I'll discuss this more on the next page. For now, just examine this code to see what's being done.

```
// Cumulate for this random vector
for (i=0; i<n_vars; i++) {
  sums[i] += xvec[i];
  for (j=0; j<=i; j++)
    covar[i*n_vars+j] += xvec[i] * xvec[j];
}
} // For all cases
```

Suppose we want to compute the covariance of a set of observed scalar random variables x and y. Let μ_x be the computed mean of x, and let μ_y be the computed mean of y. Then the "traditional" and (usually) accurate formula for their covariance is given by Equation (4.4).

$$Cov_{x,y} = \frac{1}{n}\sum_{i=1}^{n}(x_i - \mu_x)(y_i - \mu_y) \tag{4.4}$$

Unfortunately, this equation requires storage of the entire data matrix so that we can use it after computing the means. It doesn't take much manipulation to derive the mathematically equivalent Equation (4.5), which can be computed in a single pass through the dataset and hence does not require storage of the data.

$$Cov_{x,y} = \frac{1}{n}\left[\sum_{i=1}^{n}x_i y_i - \frac{1}{n}\left(\sum_{i=1}^{n}x_i\right)\left(\sum_{i=1}^{n}y_i\right)\right] \tag{4.5}$$

However, Equation (4.5) has a potentially deadly flaw when implemented on a computer. If both random variables have means whose magnitudes are large compared to their standard deviations, the subtraction in this equation will involve numbers that are both very large compared to their difference. Because computers have limited precision, many (or even most!) significant digits can be lost. Thus, Equation (4.5) should *never* be used in a general-purpose application. Either Equation (4.4) should be used, or the quite complex *online parallel formula* used. This formula is available from the Sandia National Laboratories site, among others.

But we are in luck here. The random variables are drawn from populations that have zero mean. Thus, the subtraction in Equation (4.5) is innocuous. Here is this code, without the division by n (yet).

```
// Compute n_cases times covariance
for (i=0; i<n_vars; i++) {
  for (j=0; j<=i; j++)
    covar[i*n_vars+j] -= sums[i] * sums[j] / n_cases;
  }
```

Now we convert this to a correlation matrix. The standard formula is given by Equation (4.6). Our covar matrix computation skipped the division by n in Equation (4.5), but this common factor cancels in Equation (4.6). We compute the lower triangle off-diagonal elements and then just set the diagonal to 1.0. Finally, compute the eigenvalues.

$$Corr_{x,y} = \frac{Covar_{x,y}}{\sqrt{Variance_x\,Variance_y}} = \frac{Covar_{x,y}}{\sqrt{Covar_{x,x}\,Covar_{y,y}}} \qquad (4.6)$$

```
for (i=0; i<n_vars; i++) {
  covar[i*n_vars+i] = sqrt (covar[i*n_vars+i]);
  for (j=0; j<i; j++)
    covar[i*n_vars+j] /= covar[i*n_vars+i] * covar[j*n_vars+j];
  }

for (i=0; i<n_vars; i++)  // Definition of correlation matrix
  covar[i*n_vars+i] = 1.0;

evec_rs (covar, n_vars, 0, covar, evals, workv);
return 0;
}
```

The preceding code handles the core computation. We now present the routine that coordinates multithreading of the core code. Its calling parameters are as follows:

```
int mc_evals (
   int nc,                // Number of cases
   int nv,                // Number of variables
   int mc_reps,           // Number of MC replications
   int max_threads,       // Max number of threads to use
   double fractile,       // Desired fractile, 0-1
   double *threshold      // Computed values of each eval for specified fractile
   )
```

Here are the declarations and allocation of scratch memory. If the user has specified more threads than replications, drop back the number of threads. Note that Windows imposes an upper limit on the number of threads that can run simultaneously. Specifying at most 64 should be safe for all modern versions of Windows.

```
{
   int i, j, k, ieval, ithread, n_threads, empty_slot, ret_val;
   double *covar, *evals, *workv, *all_evals;
   MC_EVALS_PARAMS mc_evals_params[MAX_THREADS];
   HANDLE threads[MAX_THREADS];

   if (mc_reps < 1)  // Silly caller
      mc_reps = 1;

   if (max_threads > mc_reps)
      max_threads = mc_reps;
/*
   Allocate memory
*/

   covar = (double *) MALLOC (nv * nv * max_threads * sizeof(double));
   evals = (double *) MALLOC (nv * max_threads * sizeof(double));
   workv = (double *) MALLOC (nv * max_threads * sizeof(double));
   all_evals = (double *) MALLOC (nv * mc_reps * sizeof(double));
```

Most parameters will be the same for all threads, so initialize them now. Notice that each thread requires its own copy of the three work areas (covar, evals, workv) so that they don't mess around with one another's private things.

```
for (ithread=0; ithread<max_threads; ithread++) {
  mc_evals_params[ithread].nc = nc; mc_evals_params[ithread].nv = nv;
  mc_evals_params[ithread].covar = covar + ithread * nv * nv;
  mc_evals_params[ithread].evals = evals + ithread * nv;
  mc_evals_params[ithread].workv = workv + ithread * nv;
  } // For all threads, initializing constant stuff
```

Get ready for and then begin the "endless" loop that handles threading. We count in n_threads the number of threads that are currently active, and ieval will count replications done. Each replication is a single thread. Each thread's handle will be stored in threads, and a NULL entry indicates that the corresponding thread is inactive
(not started or closed).

```
n_threads = 0;      // Counts threads that are active
for (i=0; i<max_threads; i++)
  threads[i] = NULL;

ieval = 0;           // Index of this trial in all_evals
empty_slot = -1;     // After full, will identify the thread that just completed
for (;;) {           // Main thread loop processes all replications
```

Compassionate programmers allow the user to interrupt potentially slow processing. It may be that a thread has completed, but the others are still running. Thus, we must crunch down the list of active threads, wait for the rest of them to finish, close them, and exit with an error code.

```
if (escape_key_pressed || user_pressed_escape ()) {
  for (i=0, k=0; i<max_threads; i++) {
    if (threads[i] != NULL)
      threads[k++] = threads[i];
    }
```

```
ret_val = WaitForMultipleObjects (k, threads, TRUE, 50000);
for (i=0; i<k; i++)
  CloseHandle (threads[i]);
ret_val = ERROR_ESCAPE;
goto FINISH;
}
```

Here is where we launch a thread if there is more work to be done. Recall that ieval counts eigenvalue-computation replications, and mc_reps is the number requested by the user. While we are initially filling the max_threads queue, empty_slot will remain at its initialized value of -1. But after the queue is filled, whenever a thread finished its work, empty_slot will be set to the position in the thread list of this now-free slot. So when we now launch a new thread, we use that just-freed slot.

We need to save in the ieval member of the parameter structure the number of this replication, as when the thread finishes, this will tell us where to put the result.

Under very rare pathological situations, Windows may not launch the thread. In this case, we must close all open threads and return an error code. Otherwise, we increment the number of active threads and the number of replications in progress or done. We know we are completely done when n_threads drops to zero: no active threads anymore.

```
if (ieval < mc_reps) {   // If there are still some to do
  if (empty_slot < 0)    // Negative while we are initially filling the queue
    k = n_threads;
  else
    k = empty_slot;
  mc_evals_params[k].ieval = ieval;     // Needed for placing final result
  threads[k] = (HANDLE) _beginthreadex (NULL, 0, evals_threaded,
                              &m c_evals_params[k], 0, NULL);
  if (threads[k] == NULL) {   // Very pathological event; should never happen
    for (i=0; i<n_threads; i++) {
      if (threads[i] != NULL)
        CloseHandle (threads[i]);
    }
```

```
      ret_val = ERROR_INSUFFICIENT_MEMORY;
      goto FINISH;
      }
    ++n_threads;
    ++ieval;
    } // if (ieval < mc_reps)

  if (n_threads == 0) // Are we done?
    break;
```

It may be that the full quota of threads are running, but there are still more replications to do. In this situation, we must pause here and wait for a thread to finish so as to free up a slot to launch another thread. The large wait time in milliseconds is fairly arbitrary; feel free to customize it. To be a conscientious programmer, we must prepare for the possibility of an error. Handle it as you see fit.

The WaitForMultipleObjects() call will return as soon as a thread finishes. When this happens, we must gather the nv array of eigenvalues computed by the thread and store them in all_evals. Note that they are stored with the replication changing fastest, which facilitates sorting later.

Finally, we preserve the index of this now free slot in the thread array, because this is the slot where the next thread will go. We close this thread now that its work is done, and we set its slot to NULL to indicate that the thread is closed. Decrement the number of active threads.

```
  if (n_threads == max_threads && ieval < mc_reps) {
    ret_val = WaitForMultipleObjects (n_threads, threads, FALSE, 500000);
    if (ret_val == WAIT_TIMEOUT || ret_val == WAIT_FAILED ||
      ret_val < 0 || ret_val >= n_threads) {
      ret_val = ERROR_INSUFFICIENT_MEMORY;
      goto FINISH;
      }

    k = mc_evals_params[ret_val].ieval;
    for (i=0; i<nv; i++)
      all_evals[i*mc_reps+k] = mc_evals_params[ret_val].evals[i];
```

```
    empty_slot = ret_val;
    CloseHandle (threads[empty_slot]);
    threads[empty_slot] = NULL;
    --n_threads;
    }
```

The last possibility is that we have no more work to start, as all replications have been launched and are completed or still running. When this time comes, we just sit here and wait until all threads have run to completion. As before, we are good little programmers and handle the possibility of an error. Exactly as we did in the prior code block, we collect the computed eigenvalues from each thread. But this time we must handle all threads in a loop, not just a single completed thread. While we are doing this, close the threads. At this point we are finished with all threaded eigenvalue computation and so break out of the "endless" loop.

```
  else if (ieval == mc_reps) {
    ret_val = WaitForMultipleObjects (n_threads, threads, TRUE, 500000);

    if (ret_val == WAIT_TIMEOUT || ret_val == WAIT_FAILED ||
      ret_val < 0 || ret_val >= n_threads) {   // Rare pathological error condition
      ret_val = ERROR_INSUFFICIENT_MEMORY;
      goto FINISH;
      }

    for (i=0; i<n_threads; i++) {   // For each thread that finished
      k = mc_evals_params[i].ieval;
      for (j=0; j<nv; j++)          // Get its computed eigenvalues
        all_evals[j*mc_reps+k] = mc_evals_params[i].evals[j];

      CloseHandle (threads[i]);
      }

    break;
    }
  } // Endless loop which threads computation of evals for all reps
```

All that's left to do is to compute the user-specified fractile (across replications) for each ordered eigenvalue. Compute k as the unbiased index and restrict it to legal values in case we have a careless user. Then, for each ordered eigenvalue, sort the replications and save the value as the threshold that will be used for choosing the number of factors to retain.

```
k = (int) (fractile * (mc_reps+1)) - 1;
if (k < 0)
   k = 0;
if (k >= mc_reps)
   k = mc_reps - 1;

for (i=0; i<nv; i++) {
   qsortd (0, mc_reps-1, all_evals + i * mc_reps);
   threshold[i] = all_evals[i*mc_reps+k];
   }

ret_val = 0;

FINISH:
   if (covar != NULL)
      FREE (covar);
   if (evals != NULL)
      FREE (evals);
   if (workv != NULL)
      FREE (workv);
   if (all_evals != NULL)
      FREE (all_evals);

   return ret_val;
}
```

Clustering Variables in a Subspace

In any application involving a large number of variables, it's nice to be able to identify sets of variables that have significant redundancy. Of course, we may be unlucky and have a situation in which the small *differences* between largely redundant variables contain the useful information. However, this is the exception. In most applications,

213

it is the redundant information that is most important; if some type of effect impacts multiple variables, it's probably important. Because dealing with fewer variables is always better, if we can identify groups of variables that have great intra-group redundancy, we may be able to eliminate many variables from consideration, focusing on a weighted average of representatives from each group, or perhaps focusing on a single factor that is highly correlated with a redundant group. Or we might just be interested in the *fact* of redundancy, garnering useful insight from it.

One popular way to identify redundant variables is to display scatterplots of variables on principal or rotated orthogonal axes. Variables that lie near one another in the plot have a form of redundancy *in the subspace defined by that pair of axes*. This method is especially popular in the field of psychology. But it has three drawbacks. First, it relies on visual impressions, which are notoriously subjective and may be difficult to see if variables crowd together. More seriously, such displays are possible in only two dimensions at a time. It is possible, even likely, that some variables will exhibit strong redundancy in some low-dimension subspace while being very independent in another, unobserved dimension. It's easy to be fooled, so arbitrary multiple-dimension consideration is much better. Last but not least, innocently flipping the sign of a variable flips its position in the plot to the opposite quadrant, destroying visual cues.

Let's develop an intuitive method for detecting redundancy of variables when this redundancy is restricted to a particular subspace. Suppose we have three unobservable, uncorrelated underlying factors: V_1, V_2, and V_3. These give rise to observed variables according to the following formulas:

$$X_1 = 1.5\ V_1 - 1.0\ V_2 + 0.7\ V_3$$
$$X_2 = 3.0\ V_1 - 2.0\ V_2 - 3.0\ V_3$$
$$X_3 = 2.0\ V_1 + 1.0\ V_2 + 1.0\ V_3$$

It should be apparent that these three observed variables do not have much redundancy with one another. X_3 has a response to V_2 opposite the other two observed variables, and X_2 has a response to V_3 opposite the others as well. Their correlation matrix would not contain values of more than moderate magnitude.

But now suppose we know (by some sort of magic, in this example!) that the unobserved third factor, V_3, is of no concern to us. Perhaps it is just noise that unjustifiably reduces correlations, and we'd rather remove its influence on our studies. We then see that X_2 is just twice X_1! In other words, these two variables are completely redundant *when considered in the context* of the two unobservable underlying factors that we believe most important. Of course, in our application, *neither alone can substitute for the knowledge gained from both of them*, because the "noise" factor V_3 impacts them quite differently. But the knowledge of this redundancy itself may give us valuable insight into the process being studied. And if we know that, in terms of the useful information, they are redundant, we may be able to replace these two variables with just their average, or their first principal component. Knowledge is power.

Continuing this intuitive development, we now are at the point of knowing that our observed variables are defined in terms of their *important* unobserved components as follows:

$$X_1 = 1.5 \ V_1 \ - \ 1.0 \ V_2$$
$$X_2 = 3.0 \ V_1 \ - \ 2.0 \ V_2$$

How can we rigorously measure the redundancy of X_1 and X_2, in this case coming up with a measure of perfect redundancy? There are many ways, but my favorite is to consider each observed variable as a vector in the space defined by the orthogonal underlying factors. Here, these vectors would be (1.5, -1.0) and (3.0, -2.0). We just compute the angle between these two vectors, agreeing that smaller angles equate to greater redundancy. In this example, the angle is zero: perfect redundancy.

Recall that the angle θ between two vectors x and y is given by Equation (4.7), in which • means dot product, and $\|.\|$ means Euclidean length.

$$\cos(\theta_{x,y}) = \frac{x \cdot y}{\|x\| \ \|y\|} \tag{4.7}$$

This gives us an alternative but equivalent way to measure redundancy: the dot product of the two vectors when their lengths have been normalized to equal one. This dot product will range from a low of -1 when the vectors point in opposite directions to a high of +1 when they are identical. This leads to another consideration: are X_1 and X_2 redundant when $X_1 = -X_2$? In most applications, we would say yes, because the sign of a variable is just dependent on some aspect of how it is measured. Another way of looking at this issue is that knowledge of X_1 provides perfect knowledge of X_2 when one is just the negative of the other. This surely fits the definition of redundancy! So we should modify our redundancy criterion in one small way: let it be the *absolute value* of the dot product of the normalized vectors.

But what are the vectors? The example just shown used values made up for this demonstration. How can we find coefficients for computing observed variables in terms of unobserved common factors? If you've been paying attention to this chapter, you will instantly know that the dominant (or perhaps all) principal components fit the bill nicely. As has been stated before, it is very often (though not always!) the case that early (large eigenvalue) principal components contain most of the useful information in a set of observed variables, while the late (small eigenvalue) components tend to be mostly irrelevant noise. Thus, we are strongly inclined to let these dominant principal components play the role of common factors.

We already saw how to compute the factor structure (correlations of factors with variables) by multiplying each eigenvector by the square root of its corresponding eigenvalue. We state without proof (available in many multivariate statistics textbooks) a rather remarkable fact: *the factor structure matrix is also the matrix of coefficients for computing the standardized observed variables from the values of the principal components (common factors).*

Thus, to compute the redundancy of a pair of variables in what is often a sensible manner, we decide how many of the dominant principal components are important. Keep that many columns of the factor structure matrix, and normalize the length of each row to unity so that we don't have to worry about the denominator in Equation (4.7). Then the redundancy of two variables in this context is the absolute value of the dot product of the corresponding two rows in this re-normalized factor structure matrix.

Now that we know how to measure the redundancy of a pair of variables, we must consider how to group variables into sets with high internal redundancy. As far as clustering algorithms go, *hierarchical clustering* is considered by many (including myself) to provide the highest quality groups. The major disadvantage of this algorithm is that its compute time is proportional to the cube of the number of items being clustered, a deadly flaw if the items number in the thousands or more. But not many practical applications have this many variables, so this is my recommended method.

The algorithm begins by letting each variable (row in the normalized factor structure matrix) define its own one-item group. Then it tests every possible pair of groups, finding the pair that is closest (most redundant; maximum absolute value of dot product). These two groups are merged into a single group, and a representative matrix row for this new group is defined. This process repeats until we get down to a single group or the redundancy measure drops to excessively small values.

When two groups are merged, there are two common methods for defining the row vector for the combined group. The easier and often slightly superior method is to just arbitrarily choose the row vector of one of the two groups being merged. A more complex and occasionally inferior method is to compute a combined centroid, a size-weighted average of the row vectors of the two merged groups. This will be discussed in more detail in the next section.

Code for Clustering Variables

The file AN_CVARS.TXT contains the core C++ code for the algorithm just described. Error checking, user escape, and other peripheral issues are omitted for clarity. The calling parameters and local variables are declared as shown next. Initialize the number of groups to be the number of variables, as we begin with each variable being its own group. We rename the number of variables from the global npred to nvars purely for clarity. The ngrp_to_print parameter lets the user control the size of the DATAMINE.LOG file's content from this operation; once the number of groups drops this low or lower, the group membership (list of variables) for each group is printed at each step. This can be very long if there are numerous variables.

```
int an_cvars (
   int n_dim,            // Number of initial dimensions to consider
   int ngrp_to_print,    // Start printing when n of groups drops this low
   int type              // Centroid versus leader method
   )
{
   int i, j, nvars, icand1, icand2, ibest1, ibest2, n_groups, *group_id, *n_in_group;
   double x, dotprod, length, best_dotprod, *centroids;
   char msg[256], msg2[256];

   n_groups = npred;  // Number of groups; initially, every variable is its own group
   nvars = npred;     // This name just makes things more clear; no other reason
```

Allocate memory. These three items have the following uses:

- group_id: For each variable, this holds the ID of the group to which it belongs

- n_in_group: For each group, this holds the number of variables in the group

- centroids: For each group, this holds the vector that defines its leader or centroid

```
group_id = (int *) MALLOC (nvars * sizeof(int));
n_in_group = (int *) MALLOC (nvars * sizeof(int));
centroids = (double *) MALLOC (nvars * n_dim * sizeof(double));
```

The following code initializes the algorithm. When we begin, each variable defines its own group, so we set the group IDs and group sizes accordingly. By normalizing each vector to unit length, we don't have to worry about the denominator in Equation (4.7).

```
for (i=0; i<nvars; i++) {
  group_id[i] = i;          // For each variable, this is the group to which it belongs
  n_in_group[i] = 1;        // Size of each group
  length = 0.0;             // Will cumulate squared length of this variable's vector
  for (j=0; j<n_dim; j++)
    length += structure[i*nvars+j] * structure[i*nvars+j];
  length = 1.0 / sqrt (length);
  for (j=0; j<n_dim; j++)   // Normalize the length of this variable's vector
    centroids[i*n_dim+j] = length * structure[i*nvars+j];
  }
```

The hierarchical clustering algorithm now begins. Each pass through the outer loop merges a single pair of groups, thus decreasing the number of groups by one. Recall that our merging criterion (measure of redundancy) is the absolute value of the dot product of the two candidate vectors. We'll keep track of the score of the best candidate pair in best_dotprod.

```
while (n_groups > 1) {
  best_dotprod = -1.0;

  // Try every pair of groups (icand1 and icand2)
  for (icand1=0; icand1<n_groups-1; icand1++) {
    for (icand2=icand1+1; icand2<n_groups; icand2++) {

      dotprod = 0.0;                    // Will cumulate for this candidate pair
      for (i=0; i<n_dim; i++)
        dotprod += centroids[icand1*n_dim+i] * centroids[icand2*n_dim+i];
      dotprod = fabs (dotprod);         // Handle symmetry
```

```
    if (dotprod > best_dotprod) {        // Keep track of the pair with best criterion
       best_dotprod = dotprod;
       ibest1 = icand1;
       ibest2 = icand2;
       }
    } // For icand2
  } // For icand1
```

For the user's information, print the results of this merger. Tiny floating-point errors may cause the computed dot product to trivially exceed its theoretical limit. This would be a problem for the acos() routine that is used to get the corresponding angle for the user, so make sure it does not happen.

```
if (best_dotprod > 1.0)   // Should never happen, but handle tiny fpt errors
   best_dotprod = 1.0;

sprintf_s (msg,
   "Merged groups %d and %d separated by %.2lf degrees; now have %d groups",
   ibest1+1, ibest2+1, acos(best_dotprod)*180.0/PI, n_groups-1);
audit (msg);     // This writes to the DATAMINE.LOG file
```

We will soon absorb the group having the larger index into the smaller. If the user requests the leader method, we just leave the "centroid" of the absorbing group alone. But if the centroid method is requested, we must compute the centroid of the merged group as a size-weighted average of the two merging groups. A more theoretically correct method would be to project the two vectors onto a plane and subdivide the angle between them on this plane. But the approximation used here is very good. Besides, I see no practical benefit to the projection method, so there is no point bothering. Remember that we must keep the vector at unit length, so normalize it.

```
if (type) { // Did the user request centroid method?
   // Recompute the (approximate) centroid of the absorbing (smaller id) group
   length = 0.0;
   for (j=0; j<n_dim; j++) {
     x = (n_in_group[ibest1] * centroids[ibest1*n_dim+j] +
          n_in_group[ibest2] * centroids[ibest2*n_dim+j]) /
          (n_in_group[ibest1] + n_in_group[ibest2]);
     centroids[ibest1*n_dim+j] = x;
     length += x * x;
     }
```

```
   length = 1.0 / sqrt (length);
   for (j=0; j<n_dim; j++)
     centroids[ibest1*n_dim+j] *= length; // The length must always be one
   } // If type is centroid (not leader)
```

Here is where we absorb the larger-index group into the smaller. The following operations are involved in this merger:

- Increment the group size of the absorbing group by the size of the absorbed group.

- Any group formerly marked as belonging to the absorbed group must be remapped to belong to the absorbing group.

- The group ID of the absorbed group is now unused, so remap all larger group IDs to be one smaller, thus filling in the gap.

- To match the "crunching down" of variable group IDs above the absorbed group, similarly move down by one slot every group size and centroid for groups above the absorbed group.

- Decrement the number of groups.

```
n_in_group[ibest1] += n_in_group[ibest2]; // Group 1 just absorbed group 2

// Remap the largest and then pull down all groups above largest.
for (i=0; i<nvars; i++) {
  if (group_id[i] == ibest2)  // If this variable was in Group 2
    group_id[i] = ibest1;     // Reclassify it as being in Group 1, the absorbing group
  if (group_id[i] > ibest2)   // Groups above absorbed group
    --group_id[i];            // Now have to fill in the hole below them
  }

for (i=ibest2+1; i<n_groups; i++) { // Crunch down stuff above absorbed group
  n_in_group[i-1] = n_in_group[i];
  for (j=0; j<n_dim; j++)
    centroids[(i-1)*n_dim+j] = centroids[i*n_dim+j];
  }
```

```
// Optionally print group membership here

--n_groups;     // We just lost a group (ibest2 was absorbed into ibest1)
} // while (n_groups > 1)
```

// Finished. Free group_id, n_in_group, and c entroids here.

Separating Individual from Common Variance

We've seen how computing the principal components of a correlation matrix, trivially derived from the eigenvectors, has many uses. We can identify the dominant directions of variance, which is usually quite revealing of the underlying structures of a set of measured variables. More importantly (in my own work, at least) is that we can then cluster variables in a dominant subspace to identify groups of redundant or nearly redundant measurements *taken in the context of the subspace, ignoring contributions from less dominant (more likely noise) subspaces.* Finally, developers willing to believe that small-eigenvalue directions have little or no relevance to their application can discard these directions and thereby create a smaller subset of new variables for their application, those based strictly on dominant components.

But when it comes to exploratory data analysis, a key first step in any research endeavor, simple principal components study suffers from several weaknesses that can seriously impede its utility. These weaknesses, discussed soon, arise from Equation (4.2) on page 187. To understand why, remember that a major goal in this preliminary data exploration is to determine if our observed variables (or some designated subset of them) are arising from some other, usually much smaller, set of unobserved (or at least unmeasured) common factors.

As an example from the medical field, we may be studying a large collection of patients and measuring the degree, presence, or absence of specific health conditions such as height, weight, various blood count statistics, frequency of headache, blood pressure, depression, and so forth. What may be difficult or impossible to observe is their unreported food consumption, illegal or unprescribed drug usage, sexual proclivities, marital happiness, and a myriad of other touchy issues. If we can at least determine the *existence* of underlying common factors driving the observed variables, we may be able to benefit from nothing more than the knowledge of their existence in terms of how they impact the observed variables. If we are lucky, we may perhaps even come up with reasonable names for these common factors, though in my experience, assigning names is of secondary importance compared to understanding their impact on the observed variables.

We can use ordinary multiple regression to invert Equation (4.2) on page 187 in order to devise Equation (4.8), which computes our observed variables **x** if we are given values for the unobserved common factors **y**.

$$\mathbf{X} = \mathbf{Ay} \tag{4.8}$$

To keep things simple in this presentation, we continue the assumption stated at the start that the observed variables that make up the **x** vector have been standardized to have zero mean and unit variance. This is not strictly required in the traditional developments. However, this assumption imposes no practical limitations of any sort, and it greatly simplifies the math that follows, as we can ignore means and scaling constants. What *is* required in this and traditional presentations is that the **y** vector components, the unobserved common factors, has zero mean and unit variance. If you want more rigorous mathematics instead of the simplified versions in this text, you can easily find detailed presentations all over the Internet and in statistics references.

Surprisingly to many, it turns out that the **A** matrix of Equation (4.8) is just the factor structure matrix we discussed on page 189. In other words, the matrix of correlations between the observed variables and the unobserved common factors is also the regression matrix that lets us (if we were able!) compute the observed variables from the unobserved common factors. (Wow!) If the correlation matrix of the observed data is full rank (no perfect colinearity), and if we keep all eigenvectors, this computation is exact. Otherwise, the computed values of **x** from Equation (4.8) are least- squares approximations.

We have one last interesting tidbit to present before getting on with the main topics of this section: a serious problem with principal components when used for initial data exploration, and a solution for this problem. Recall that we are designating **R** as the correlation matrix of the raw data **x**. Another fundamental equation from principal components is that we can reproduce this correlation matrix from the factor structure (often called the *factor loading* matrix when used in this regression context). This is shown in Equation (4.9).

$$\mathbf{R} = \mathbf{AA}' \tag{4.9}$$

If **A** contains all factors (a square matrix), the reproduction is exact. If some columns of **A** have been removed (some principal components discarded as unimportant), then the reproduction is an approximation.

Pant, pant. At long last we are ready to discuss the data-exploration issues with Equation (4.2) on page 187 and the two equations just shown. The heart of the problem is that the observed-to-factor equation, (4.2), and the factor-to-observed equation, (4.8), are nothing more than transformations. They map one set of variables to another set of variables. And Equation (4.9) is almost trivial, showing how under the principal components model, the correlation matrix of the data is explained by nothing more than the product of the factor loading matrix with its own transpose. This formulation does have a certain elegant simplicity, but we would much rather have a more general, powerful model for expressing the impact of unobservable common factors on our observed variables.

In particular, in addition to the variance that is attributable to the common factors, *we would like to be able to account for any degree of variance in each observed variable that is **unique** to that variable.* It is a severe limitation to require that *all* of the variation we see in an observed variable be attributable to common factors. We want to assume the existence of unique variance as well. This unique variance may be valid information not related to the common factors, or it may just be random noise. Regardless, requiring that the hypothetical common factors be able to account for all variance in all observed variables is a significant impediment to easy interpretation of numerical results. It forces the computed **A** matrix to conform to unreasonable expectations. Noise happens, and if we pretend it doesn't, we pay a price.

So let's slightly modify the model. Equation (4.8) shows that our observed variables are just linear combinations of the unobserved factors. We make one seemingly trivial change, and in return we get enormously increased power. Just let the observed vector **x** also include an error vector ε, as shown in Equation (4.10).

$$\mathbf{X} = \mathbf{A}\mathbf{y} + \varepsilon \tag{4.10}$$

We make the innocuous assumption that the error vector follows a multivariate normal distribution with mean zero, and the covariance matrix of this error vector is diagonal. In other words, the errors for the observed variables are uncorrelated, and their variances need not be equal. These variances are traditionally designated by the Greek letter Psi (Ψ).

Before venturing any further into the mathematics of what is traditionally called *maximum likelihood factor analysis*, let's take a look at a motivational example of what the inclusion of this little error term can do for us. I created ten independent random variables called **RAND0** through **RAND9**. I then defined three new random variables in terms of several of them, with the idea that RAND1 through RAND4 can serve as both unobserved common factors and observed variables:

> **SUM12 = RAND1 + RAND2**
>
> **SUM34 = RAND3 + RAND4**
>
> **SUM1234 = SUM12 + SUM34**

Look at the two tables on the next page, which arise from keeping the four most dominant eigenvectors of this dataset's correlation matrix. And you might want to review the definition of *communality* given on page 193. Communality is the sum of the squares of the factor structure for that variable, and it expresses the fraction of the variance of each observed variable that is explained by the retained factors. The observed variables have been standardized to unit variance, so one minus the communality of a variable can be loosely interpreted as the unexplained variance, the variance of an observed variable not attributable to the common factors that the user retained. This is loosely analogous to *Psi*, the variance of the error term just discussed.

The topmost of these two tables is a principal components analysis, which disallows explicit inclusion of unexplained variance. Psi can only be roughly inferred as one minus the communality, a clumsy and often inaccurate approach. (For example, with RAND0, $0.8056 = 1 - 0.0122^2 - 0.0066^2 - 0.3741^2 - 0.2329^2$.) The three sum variables (SUM12, SUM34, SUM1234) in this top table have small inferred unexplained variance, as expected since they have much in common with other observed variables. The four variables that go into these sums, RAND1 through RAND4, also have smallish unexplained variance, while the other variables are larger.

But compare this to the bottom table, which is the result of the factor analysis procedures to be described in this section. Now the distinction between observed variables that have common ancestry and those that do not is abundantly clear. The seven variables that share underlying driving forces have an independent-variance measure (Psi) of zero, while the variables that have nothing in common are shown to be nearly 100 percent independent. The difference in interpretability is profound.

Initial evals, cumulative pct, Psi, and loadings

	Eigenvalue	2.983	2.019	1.068	1.044
	Cumulative	22.945	38.474	46.688	54.718

Initial Psi

RAND0	0.8056	-0.0122	0.0066	0.3741	0.2329
RAND1	0.2052	0.4851	0.4980	-0.5263	-0.1858
RAND2	0.2050	0.4664	0.5247	0.5167	0.1873
RAND3	0.3942	0.5149	-0.4958	0.1883	-0.2437
RAND4	0.4028	0.5222	-0.4822	-0.1692	0.2518
RAND5	0.6796	0.0086	0.0043	-0.5326	0.1917
RAND6	0.7785	0.0082	0.0479	0.0341	0.4669
RAND7	0.8039	-0.0287	0.0109	-0.0742	0.4355
RAND8	0.7791	0.0019	0.0045	-0.0287	0.4691
RAND9	0.8299	0.0093	0.0943	0.1684	-0.3643
SUM12	0.0017	0.6805	0.7315	-0.0065	0.0013
SUM1234	0.0010	0.9997	0.0205	0.0054	0.0045
SUM34	0.0011	0.7270	-0.6856	0.0138	0.0051

Final factor variances, Psi, and factor loadings

	Squared length	2.982	2.010	0.844	0.736

Final Psi

RAND0	0.9991	-0.0080	0.0039	0.0255	0.0012
RAND1	0.0000	0.4861	0.4965	-0.6099	-0.2400
RAND2	0.0000	0.4654	0.5262	0.6003	0.2415
RAND3	0.0000	0.5174	-0.4915	0.2427	-0.5519
RAND4	0.0000	0.5196	-0.4866	-0.2238	0.5611
RAND5	0.9985	0.0058	0.0022	-0.0346	-0.0009
RAND6	0.9988	0.0055	0.0251	0.0191	0.0106
RAND7	0.9989	-0.0193	0.0044	-0.0083	0.0219
RAND8	0.9998	0.0014	0.0029	-0.0035	0.0122
RAND9	0.9975	0.0064	0.0483	0.0096	-0.0049
SUM12	0.0000	0.6805	0.7315	-0.0065	0.0012
SUM1234	0.0000	0.9997	0.0205	0.0054	0.0045
SUM34	0.0000	0.7270	-0.6857	0.0138	0.0051

Astute readers familiar with factor analysis will notice a peculiarity about the second table: in traditional factor analysis, the sum of squares of the loadings in each row, plus the Psi for that row, add up to the variance of the observed variable of that row. (This identity may become clearer in a moment when we discuss the upcoming Equation (4.11).) Because our observed variables have been standardized, this sum should be 1.0, but for several rows the sum doesn't quite make it. This is because there is some perfect colinearity in the dataset; the SUM variables are exact functions of some of the RAND variables. In traditional factor analysis, such colinearity is forbidden. But in the algorithm that I use, colinearity usually does not cause numerical difficulties, so I allow it, especially since the results of this loose algorithm can make colinearities obvious, as happened in that contrived example. If you have no idea what this paragraph just said, don't worry about it; just be aware that if your data does contain any perfect colinearity, results may be somewhat compromised, but the colinearity will likely be revealed and thereby made easy to eliminate before further study is made!

Now that we're nicely motivated, let's proceed with an overview of the mathematics of maximum likelihood factor analysis. As is my usual practice, I keep the mathematical detail limited to the bare minimum needed to gain an intuitive understanding of what's going on and to understand the computer code that will follow. If you feel cheated of rigor, you will have no trouble finding what you desire on the Internet and any of the numerous textbooks on the subject. Later, when the code is presented, I'll mention two particularly useful publications.

Equation (4.8) on page 222 shows how, in the principal components model, the observed variables are produced by the unobserved factors. This led to Equation (4.9) showing how the correlation matrix of the observed variables relates to the loadings. Now we extend this idea to include the unexplained-variance term. In this more general model, we can't call the covariance matrix of the observed variables a correlation matrix, although the analogy is strong. Thus, instead of referring to it as **R**, we'll follow the tradition of using the Greek letter sigma (Σ) to designate the covariance matrix of the observed variables, **x**. As mentioned earlier, the covariance matrix Ψ of the ε term is diagonal, with the individual variances on the diagonal. Then, when our model is given by Equation (4.10) on page 223, the analog of Equation (4.9) on page 222 is given by Equation (4.11).

$$\Sigma = \mathbf{AA}' + \Psi \tag{4.11}$$

This equation should satisfy our intuition, because it says that the covariance of a model that includes unique variance for each measured variable is just the covariance created by the common-factor loadings plus the unique variances.

In the simple principal components model (no unique variances), estimating the **A** matrix is trivial; it's just the eigenvectors, each multiplied by the square root of its corresponding eigenvalue. But when we include unique variance terms, things become a lot messier. No direct solution is possible. The most common (and likely best) approach is to find **A** and Ψ, which maximize the normal-distribution likelihood function associated with this model.

If there are n cases, the log likelihood function is given by Equation (4.12), in which |.| means the determinant of the matrix, $tr(.)$ means the trace (sum of diagonal elements), and **S** is the sample covariance matrix (which in our context is also the sample correlation matrix, because the observed variables are standardized). Also, Σ is defined by Equation (4.11).

$$l(\mathbf{A}, \Psi) = -\frac{n}{2}\left[\ln|\Sigma| + tr\left(\Sigma^{-1}\mathbf{S}\right)\right] \tag{4.12}$$

For the remainder of this discussion of maximum likelihood factor analysis, including the code presented later, we'll often be mentioning two constants in the application, so we'll give them names now. There are npred measured variables. (This name comes from the fact that the variables are most likely predictor candidates in the application.) And we are assuming that there are n_dim unobserved common factors. The developer is responsible for coming up with a reasonable guess for n_dim, although later we'll discuss how this guess can be made somewhat intelligently. Naturally, n_dim <= npred, and n_dim will be much less than npred in nearly any practical application.

This dimensionality difference inspires an important observation about the log likelihood function, Equation (4.12). The Σ matrix is npred square, and in many applications npred will be quite large. In some of my applications, npred might be on the order of 100 variables, or even 1000, while n_dim might be 5 to 10 or so. Equation (4.12) involves inverting and finding the determinant of a potentially gigantic matrix, not a trivial undertaking.

Luckily, the definition of Σ given by Equation (4.11) lets us write its determinant and inverse in a way that is a *lot* faster to compute. Don't even think about using the naive version of Equation (4.12). The required quantities are given in Equations (4.13) and (4.14), respectively. The derivation of these fierce identities can be found in several sources, the most detailed (I believe) being Chapter 4 of *Factor Analysis as a Statistical Method, 2nd Ed* by Lawley and Maxwell.

$$|\Sigma| = |\Psi| |\mathbf{I} + \mathbf{A}' \Psi^{-1} \mathbf{A}| \tag{4.13}$$

$$\Sigma^{-1} = \Psi^{-1} - \Psi^{-1} \mathbf{A} \left(\mathbf{I} + \mathbf{A}' \Psi^{-1} \mathbf{A} \right)^{-1} \mathbf{A}' \Psi^{-1} \tag{4.14}$$

Because Ψ is a diagonal matrix, its inverse is also a diagonal matrix containing the reciprocals of the diagonal elements of Ψ. That's a trivial operation. And the key is that the only general matrix that must be inverted is n_dim square, which in nearly all practical operations will be a whole lot faster than inverting an npred square matrix. As for the determinant, Equation (4.13), both terms are easy. The determinant of Ψ is just the product of its diagonal elements, and the general matrix whose determinant is needed is the same matrix that has to be inverted for Equation (4.14). For those who were sleeping that day in linear algebra class, know that the determinant of a matrix is trivial to compute as part of the inversion process.

Log Likelihood the Slow, Definitional Way

In this short section I'll present code for directly using Equation (4.12) to compute the log likelihood function (except for the factor of $n/2$, which is constant and would be just a waste of computer time). No sane programmer would use this method, as it involves inversion of a potentially gigantic matrix. However, it is instructive and simple and therefore worthy of a quick treatment.

In this code, we concatenate the Ψ diagonal matrix containing npred parameters with the npred by n_dim matrix of factor loadings, $\mathbf{A}$, into a single vector that we will call theta (θ). This greatly simplifies some optimization code that we'll see later. So the first step here is to split them apart into PSI and A. Then we use Equation (4.11) to compute Σ in TEMPmat1.

```
double AnalyzeFactorChild::log_lik (double *theta)
{
  int i, j, k;
  double sum, det, *PSI, *A;
```

```
PSI = theta;
A = theta + npred;

/*

  Sigma inverse = (Psi + A A') inverse
  Determinant of Sigma
*/

  for (i=0; i<npred; i++) {
    for (j=0; j<npred; j++) {
      sum = 0.0;
      for (k=0; k<n_dim; k++)
        sum += A[i*n_dim+k] * A[j*n_dim+k];
      TEMPmat1[i*npred+j] = sum;      // A A'
      }
    TEMPmat1[i*npred+i] += PSI[i];     // This completes Equation (4.11)
    }
```

Given the safety precautions in the implementation, it would be highly unusual for Σ to be singular, but if our inversion routine reports this unfortunate event, we return such a horrendous log likelihood that this problematic search region will be abandoned by the optimization algorithm. Our inversion routine (the source code is in INVERT.CPP) computes the determinant of the matrix as an efficient byproduct of inversion. Then we trivially complete Equation (4.12). Because we need only the trace of the matrix product, we avoid computing off-diagonal elements. Recall that covar is symmetric, so we can access elements in either direction. The direction used here is somewhat faster on some compilers.

```
  k = invert (npred, TEMPmat1, TEMPmat2, &det, invert_rwork, invert_iwork);
  if (k)
    return -1.e60;

/*

  Trace of above times covar
*/
```

```
  sum = 0.0;
  for (i=0; i<npred; i++) {
    for (k=0; k<npred; k++)
      sum += TEMPmat2[i*npred+k] * covar[i*npred+k];
    }

  return -log(det) - sum;
}
```

Log Likelihood the Fast, Intelligent Way

This method, which is mathematically identical to the direct method just shown, can be orders of magnitude faster than the direct method because of one reason only: the matrix that we must invert will almost always be much smaller than that in the direct method. We still use the same definition of log likelihood, Equation (4.12), but we compute Σ^{-1} and the determinant more efficiently, using Equations (4.13) and (4.14). Here is the code:

```
double AnalyzeFactorChild::log_lik_fast (double *theta)
{
  int i, j, k;
  double sum, det, *PSI, *A;

  PSI = theta;
  A = theta + npred;

/*
  We compute the inverse and determinant of sigma using the fast method
*/

  // Compute F = PsiInverse A, a component of Equations 4.13 and 4.14 on Page 228

  for (i=0; i<npred; i++) {
    for (j=0; j<n_dim; j++)
      Fmat[i*n_dim+j] = Amat[i*n_dim+j] / PSIvec[i];
    }
```

```
// (A'F + I) completes the n_dim by n_dim matrix which we must invert

for (i=0; i<n_dim; i++) {
  for (j=0; j<n_dim; j++) {
    sum = 0.0;
    for (k=0; k<npred; k++)
      sum += Amat[k*n_dim+i] * Fmat[k*n_dim+j];
    TEMPmat1[i*n_dim+j] = sum;        // This is A' F
    }
  TEMPmat1[i*n_dim+i] += 1.0;         // Add in the identity matrix
  }

// Invert the matrix; in extremely rare case that it is singular, return horrid log likelihood
// This also gives us the determinant we will need later

k = invert (n_dim, TEMPmat1, TEMPmat2, &det, invert_rwork, invert_iwork);
if (k)
  return -1.e60;

// Premultiply that by F = PsiInverse A to continue Equation 4.14

for (i=0; i<npred; i++) {
  for (j=0; j<n_dim; j++) {
    sum = 0.0;
    for (k=0; k<n_dim; k++)
      sum += Fmat[i*n_dim+k] * TEMPmat2[k*n_dim+j];
    TEMPmat1[i*n_dim+j] = sum;
    }
  }

// Postmultiply that by F Transpose and simultaneously subtract it from Psi Inverse
// This completes Equation 4.14, giving us the inverse of Sigma

for (i=0; i<npred; i++) {
  for (j=0; j<npred; j++) {
    if (i == j)
      sum = 1.0 / PSIvec[i];   // Psi Inverse; we subtract from this
```

```
      else
        sum = 0.0;
      for (k=0; k<n_dim; k++)
        sum -= TEMPmat1[i*n_dim+k] * Fmat[j*n_dim+k];
      TEMPmat2[i*npred+j] = sum;
      }
   }
```

```
// The rest of this code is identical to the slow method, just Equation 4.12 without n/2
// Compute the trace of sigma-inverse times covar

sum = 0.0;
for (i=0; i<npred; i++) {
  for (k=0; k<npred; k++)
    sum += TEMPmat2[i*npred+k] * covar[i*npred+k];
  }

// Finish computation of the determinant of Sigma

for (i=0; i<npred; i++)
  det *= PSIvec[i];

return -log(det) - sum;
}
```

The Basic Expectation Maximization Algorithm

Even with the simplifications just presented, direct numerical maximization of
Equation (4.12) is much too slow to be practical. With the discovery some years ago
of a wide family of optimization algorithms called *expectation maximization*,
we suddenly had a method of maximizing the log likelihood with an iterative algorithm
that, under very reasonable conditions, is guaranteed to converge to a global maximum
(there are an infinite number of them). Full theoretical derivation of this algorithm is
far beyond the scope of this text. However, we will present the key equations for an
efficient implementation of this algorithm, which is a core component of the faster
method shown in the next section. The clever sequence of operations given here is taken
from the very helpful paper "ML Estimation for Factor Analysis: EM or Non-EM?" by
J. H. Zhao, Philip L. H. Yu, and Qibao Jiang. This paper can be downloaded for free from

several sites on the Internet; a quick search will find it. If you have no luck, contact me at my website email address and I'll send you a PDF.

The algorithm begins by using ordinary principal components to find starting estimates for **A** and Ψ:

1. Compute **S**, the covariance matrix of the observed variables. Because we standardize these variables, this is also their correlation matrix, although standardization is not required for the general form of the algorithm. However, standardization aids numerical stability, so I always do it.

2. Compute the starting estimate of **A** by keeping the n_dim dominant eigenvectors of the covariance matrix and multiplying each eigenvector by the square root of its corresponding eigenvalue. Thus, we have $\mathbf{A_0}$ as an npred by the n_dim matrix.

3. Compute the starting estimate of Ψ by subtracting the variance of each variable implied by **AA**′ from the actual covariance. Look back at Equation (4.11) on page 226. Assume for this starting approximation that $\Sigma = \mathbf{S}$ and solve for Ψ, as shown in Equation (4.15).

$$\Psi_0 = diag\left(\mathbf{S} - \mathbf{AA}'\right) \tag{4.15}$$

The basic expectation-maximization (EM) algorithm then iterates as shown next. Each iteration increases the log likelihood function, although in practice convergence can sometimes be excruciatingly slow.

$$\mathbf{F} = \Psi_t^{-1} \mathbf{A}_t \tag{4.16}$$

$$\mathbf{G} = \mathbf{SF} \tag{4.17}$$

$$\mathbf{H} = \mathbf{G}\left(\mathbf{I} + \mathbf{A}_t' \mathbf{F}\right)^{-1} \tag{4.18}$$

$$\mathbf{A}_{t+1} = \mathbf{G}\left(\mathbf{I} + \mathbf{H}'\mathbf{F}\right)^{-1} \tag{4.19}$$

$$\Psi_{t+1} = diag\left[\mathbf{S} - \mathbf{HA}_{t+1}'\right] \tag{4.20}$$

There are several issues to consider when programming the basic EM algorithm:

- Equation (4.16) implies that the independent variances (the diagonal of Psi) must be positive, lest we divide by zero. This can be imposed by checking the new values computed by Equation (4.20) and resetting them slightly above zero if necessary.

- This diddling with Psi ruins the guaranty of convergence, although in practice, as long as you let them get very close to zero, this should not be a problem. Nevertheless, a responsible programmer takes into account that the algorithm could fall into an endless loop of EM driving Psi below the limit and then the program pushing it back up again. Users hate endless loops.

- Equations (4.18) and (4.19) involve inversion of a matrix that could, in rare pathological cases, be singular. Make sure you use an inversion routine that reports singularity and gracefully abort if it happens. It is extremely rare, but we do care, do we not?

Code for Basic Expectation Maximization

The class function that implements the algorithm shown in the prior section can be found in the file AN_FACTOR.TXT. Here we present it, along with a discussion of salient points as needed. The full context of this routine will appear later, but because it is straightforward and all variables are clearly named to correspond to the equations, I'll present it here, immediately after the algorithm. Memory allocations for the many arrays can be found on page 248.

```
int AnalyzeFactorChild::EMstep ()
{
   int i, j, k;
   double sum;
/*
   Compute F = PsiInverse A which is Equation (4.16)
   We trust that we have never let PSIvec drop to a computational zero.
*/
```

```
for (i=0; i<npred; i++) {
  for (j=0; j<n_dim; j++)
    Fmat[i*n_dim+j] = Amat[i*n_dim+j] / PSIvec[i];
  }

/*
  Compute G = covar F which is Equation (4.17)
  Recall that S in the equation is the covariance (correlation) matrix 'covar'
*/

for (i=0; i<npred; i++) {
  for (j=0; j<n_dim; j++) {
    sum = 0.0;
    for (k=0; k<npred; k++)
      sum += covar[i*npred+k] * Fmat[k*n_dim+j];
    Gmat[i*n_dim+j] = sum;
    }
  }

/*
  Compute H in multiple steps for Equation (4.18)
*/
  // (A'F + I) Inverse
  for (i=0; i<n_dim; i++) {
    for (j=0; j<n_dim; j++) {
      sum = 0.0;
      for (k=0; k<npred; k++)
        sum += Amat[k*n_dim+i] * Fmat[k*n_dim+j];
      TEMPmat1[i*n_dim+j] = sum;  // A' F
      }
    TEMPmat1[i*n_dim+i] += 1.0; // This is where we add in the identity matrix
    }

  k = invert (n_dim, TEMPmat1, TEMPmat2, &sum, invert_rwork, invert_iwork);
  if (k)  // This would be an extremely rare pathological event that requires abort
    return 1;
```

```
// G times above finishes Equation (4.18)

for (i=0; i<npred; i++) {
  for (j=0; j<n_dim; j++) {
    sum = 0.0;
    for (k=0; k<n_dim; k++)
      sum += Gmat[i*n_dim+k] * TEMPmat2[k*n_dim+j];
    Hmat[i*n_dim+j] = sum;
    }
  }

/*
  Update A in several steps for Equation (4.19)
*/
// (H'F + I) Inverse
for (i=0; i<n_dim; i++) {
  for (j=0; j<n_dim; j++) {
    sum = 0.0;
    for (k=0; k<npred; k++)
      sum += Hmat[k*n_dim+i] * Fmat[k*n_dim+j];
    TEMPmat1[i*n_dim+j] = sum; // H' F
    }
  TEMPmat1[i*n_dim+i] += 1.0; // This is where we add in the identity matrix
  }

k = invert (n_dim, TEMPmat1, TEMPmat2, &sum, invert_rwork, invert_iwork);
if (k)  // This would be an extremely rare pathological event that requires abort
  return 1;

// G times above completes Equation (4.19)

for (i=0; i<npred; i++) {
  for (j=0; j<n_dim; j++) {
    sum = 0.0;
    for (k=0; k<n_dim; k++)
      sum += Gmat[i*n_dim+k] * TEMPmat2[k*n_dim+j];
    Amat[i*n_dim+j] = sum;
    }
  }
```

```
/*

   Update Psi = diag (covar - H A') which is Equation (4.20)
   We limit it away from zero, because inversion of matrices becomes unstable as Psi gets small.
   The consequence of this limiting is that, theoretically at least, increase of log likelihood is no longer
   guaranteed. In practice, I think decrease would be nearly impossible.
   Nonetheless, you must prepare for this possibility when this routine is invoked.
*/

   for (i=0; i<npred; i++) {
      sum = covar[i*npred+i];
      for (k=0; k<n_dim; k++)
         sum -= Hmat[i*n_dim+k] * Amat[i*n_dim+k];
      if (sum < 1.e-6)        // We must keep Psi away from zero to avoid fpt issues
         sum = 1.e-6;
      if (sum > 1.0 - 1.e-6)  // Not usual; my own restriction due to standardization
         sum = 1.0 - 1.e-6;
      PSIvec[i] = sum;
      }

   return 0;     // Tells caller that all is good in the world
}
```

Accelerating the EM Algorithm

Because the EM algorithm just presented can often suffer from slow convergence
(a tendency to zigzag back and forth across the parameter domain), great effort has
gone into finding ways to speed convergence. An Internet search will reveal a vast array
of methods. I've studied most of them and done considerable experimentation. In my
opinion, the best (fastest and most reliable convergence) has been named *DECME-2s*
by its authors. Theoretical details can be found in the manuscript *The Dynamic ECME
Algorithm* by Yunxiao He (Yale University) and Chuanhai Liu (Purdue University).
It should be easy to find on the Internet. If you have no luck, send me an e-mail at my
web site and I'll email you a PDF.

Here is an overview of how this acceleration algorithm works. We iterate two
very different optimization steps; this iteration will be discussed later, when the code
is presented. One step is the EM algorithm just shown. The other step is quadratic

optimization, which is the subject of this section. We alternate them in a loop until convergence is obtained. Note that the loading matrix is unique only up to orthogonal rotation, so there is an infinite number of equivalent global maxima.

As was mentioned in the log likelihood code, this presentation is easier if we concatenate the Ψ diagonal matrix containing npred parameters with the npred by n_dim matrix of factor loadings, $\mathbf{A}$, into a single vector that we will call theta (θ). We will roughly follow the presentation in the He and Liu paper but change a few bits of notation in a way that improves readability, at the minor cost of some rigorous notational correctness. Any such compromises are purely notational and in the spirit of specializing in the current application, and they do not damage mathematical correctness.

Suppose we have been iterating long enough to have evaluated the log likelihood at three different points. The most current point (parameter set) is theta_t (θ_t) with computed log likelihood LL_t. The immediately prior point is theta_tm1 (θ_{t-1}) with computed log likelihood LL_tm1, and the point before that is theta_tm2 (θ_{t-2}) with computed log likelihood LL_tm2. Also suppose we have just completed an EM step as described in the prior two sections. We now embark on what is called a *QUAD* step.

The idea behind a QUAD step is that, especially when in the vicinity of a global maximum, the log likelihood function tends to become approximately quadratic. There are any number of ways we could take advantage of this fact. We could pick any single parameter, or combination of parameters defining a direction, fit a parabola, and find the maximum of this parabola as an ideally better point. Or we could use two parameters or directions or three or however many we wish, fit a quadratic surface, and find the maximum of this surface. Of course, the more directions we employ, the more free parameters must be estimated for the quadratic surface fit and hence the more (very expensive!) evaluations of the log likelihood function nearby are needed. He and Liu compromise on using two directions.

Which two directions are best? The direction taken by the just-completed EM step, which is θ_t - θ_{t-1}, certainly is reasonable; perhaps the EM step was on the right track with the direction but stepped a little too far or not quite far enough. Much study indicates that a major weakness of EM is that it zigzags back and forth, closely retracing prior movements like a sailboat tacking into the wind, or a switchback path up a mountainside. This inspires us to use θ_t - θ_{t-2} as the other direction for the quadratic fit. It is likely to be fairly orthogonal to the first direction yet lie on a good plane in regard to most parameters. Thus, it is reasonable to approximate the log likelihood function in the vicinity of θ_t by Equation (4.21), which is the actual log likelihood function when it is restricted to the two directions just described.

$$f(x,y)=l\left[\theta_t+x(\theta_t-\theta_{t-1})+y(\theta_t-\theta_{t-2})\right]$$ (4.21)

We then approximate this function with the quadratic function f^* shown in Equation (4.22). **H** is the two-by-two symmetric matrix of the second-order coefficients, with constants c and d on the diagonal, and e off-diagonal.

$$f^*(x,y)=f_0+(x,y)(a,b)'+(x,y)\mathbf{H}(x,y)'$$ (4.22)

This quadratic approximation has six free parameters (f_0, a, b, c, d, e), so we need six independent points at which the log likelihood is evaluated. For maximum numerical accuracy, they should be well separated and in the vicinity of θ_t. As was stated at the beginning of this section, we already have three such points (θ_t, θ_{t-1}, and θ_{t-2}) that define the plane in which we are operating. The logical choice for one of these would be to shoot past θ_t in the θ_t - θ_{t-1} direction, going the same distance, thus placing θ_t midway between θ_{t-1} and the new point. We could do the same with θ_{t-2}, again having θ_t be midway between θ_{t-2} and the new point. The sixth and final point does not have such nice symmetry, but the logical choice would be to move past θ_t in the direction and distance of θ_{t-1} - θ_{t-2}. There is no guarantee that these six points are spaced well enough apart to ensure numerical accuracy, and we should check on this, but in most cases they will be fine.

There is a complication: the individual, unique variances on the diagonal of Ψ cannot fall to zero, lest Equation (4.16) on page 233 perform the unthinkable. In fact, they cannot even get very close to zero, as this would introduce numerical instability all over the place. Moreover, my own version of the maximum likelihood algorithm imposes the additional restriction that the unique variances cannot get excessively close to one. *This is not standard practice.* The general algorithm does not require that the observed variables be standardized. As a consequence, there is no upper bound on the unique variances. But my implementation standardizes the variables, so a variance in excess of one makes no sense. It still may happen that the **A** matrix of factor loadings can imply variance greater than one, but in practice this tends to not happen, and even if it were to happen, the practical implication for data exploration are inconsequential, so no restrictions are placed on **A**. But standardization and enforcement of a 0-1 range for the unique variance makes interpreting these very important parameters easy. This is the justification for my modification of the usual algorithm. If you don't like it, refraining from standardization and removing upper bounds in the few places they occur in the code is trivially easy.

This 0-1 restriction means that we can't just automatically jump past θ_t as we find the three new points that complete the set of six. We have to make sure that we do not jump past a limit of zero or one. The easiest way to do this is to limit the jump size by letting the new point be θ_t plus a multiplier times the distance and direction defining the jump. Ideally, this multiplier will be one, which will leave θ_t centered as discussed earlier. But if this jump would take us outside a limit, we lower the multiplier as needed. In particular, we define the three new points as follows:

$$\xi_1 = \theta_t + \alpha_1 \left(\theta_t - \theta_{t-1} \right) \tag{4.23}$$

$$\xi_2 = \theta_t + \alpha_2 \left(\theta_t - \theta_{t-2} \right) \tag{4.24}$$

$$\xi_3 = \theta_t + \alpha_3 \left(\theta_{t-1} - \theta_{t-2} \right) \tag{4.25}$$

In each of these three cases, for the sake of good spacing we let á be 1.0 if possible, but less if needed to stay inside the limit. If it turns out that á needs to be tiny in order to stay inside the limit, there's no point in continuing, because the points will be too close; computation of the quadratic fit coefficients will be ill-conditioned.

We already know the log likelihood of θ_t, θ_{t-1}, and θ_{t-2}. We compute the log likelihood of each of the three new points. The constant f_0 in Equation (4.22) would clearly best be $l(\theta_t)$ so that the function is centered there. The remaining five coefficients are computed as shown here:

$$a = \frac{l(\xi_1) - l(\theta_t) - \alpha_1^2 \left[l(\theta_{t-1}) - l(\theta_t) \right]}{\alpha_1 + \alpha_1^2} \tag{4.26}$$

$$b = \frac{l(\xi_1) - l(\theta_t) - \alpha_2^2 \left[l(\theta_{t-2}) - l(\theta_t) \right]}{\alpha_2 + \alpha_2^2} \tag{4.27}$$

$$c = l(\theta_{t-1}) - l(\theta_t) + a \tag{4.28}$$

$$d = l(\theta_{t-2}) - l(\theta_t) + b \tag{4.29}$$

$$e = -\frac{l(\xi_3) - l(\theta_t) - (a-b)\alpha_3 - (c+d)\alpha_3^2}{2\alpha_3^2} \tag{4.30}$$

The quadratic function expressed in Equation (4.22) on page 239 has a zero gradient at the (x,y) point given by Equation (4.31). This will usually be its maximum, although it will often be a saddle point. Only under rare pathological conditions will it be a minimum. Note that in the He and Liu paper cited earlier, they accidentally omit the minus sign.

$$(x,y) = -\frac{1}{2}(a,b)\mathbf{H}^{-1} \tag{4.31}$$

Once we have computed the a-e coefficients and found the stationary point of the quadratic approximation by using Equation (4.31), we are almost ready to test that point to see if it is an improvement. (It's not unusual for the improvement to be huge!)

But as when we found the three extra ξ points, we have to worry about remaining inside our 0-1 interval for the unique variances. We handle the problem in essentially the same way, by moving in the (x,y) direction from θ_t as far as we can if we cannot get all the way to (x,y). This is expressed in Equation (4.32).

$$\xi_4 = \theta_t + \alpha_4\left[x(\theta_t - \theta_{t-1})\right] + y(\theta_t - \theta_{t-2}) \tag{4.32}$$

As we did with the three extra points, we try to let $\alpha_4=1$, in which case ξ_4 is exactly at the stationary point of the quadratic fit. But if this point lies outside the permissible range of 0-1 for any unique variance, we shrink α_4 as needed to bring it into the fold.

To finish, we select whichever of these seven points has the greatest log likelihood.

Code for Quadratic Acceleration with DECME-2s

Much of the code for the algorithm of the previous section is just tedious repetition. The complete code, minus most error checking that depends on the implementation, can be found in AN_FACTOR.TXT. The presentation here will skip over a few sections that are redundant to prior code blocks. Because some coding issues are tricky but important, explanatory text will be interspersed with the code. Memory allocations for the many arrays can be found on page 248.

We begin with some basic initialization. The number of parameters is the number of unique variances plus the number of factor loadings. When this routine is called, theta_t contains the most recent parameters, those just computed by EMstep(), and LL_t is their log likelihood. (The tm1 and tm2 earlier points and their log likelihoods are also available.) These may end up being the best we've got because of QUADstep() failing to cause any improvement. So initialize the best and return value to handle this possibility. Finally, initialize a flag to indicate if any ill-condition situations arise.

```
void AnalyzeFactorChild::QUADstep (double *LLret)
{
  int i, nparams, ill_conditioned;
  double direc, alim, alim1, alim2, alim3, alim4;
  double x, y, det, aa, bb, cc, dd, ee, cci, ddi, eei;

  nparams = npred + npred * n_dim;        // Psi, A
  *LLret = LL_t;                          // We return log likelihood here
  memcpy (best_theta, theta_t, nparams * sizeof(double)); // Keep track of best here

  ill_conditioned = 0;                    // Will be set if trouble happens
```

We now have to compute the three new points, those based on $\theta_t - \theta_{t-1}$, $\theta_t - \theta_{t-2}$, and $\theta_{t-1} - \theta_{t-2}$. We'll present only the first, as the second and third are nearly identical. The following code computes α_1 (alim1 in the code) in Equation (4.23) on page 240.

```
alim1 = 1.0;          // This is the ideal value, as it creates symmetric spacing
for (i=0; i<npred; i++) {
  direc = theta_t[i] - theta_tm1[i];
  if (direc > 0.0)
    alim = (1.0 - 1.e-5 - theta_t[i]) / direc;
  else if (direc < 0.0)
    alim = (1.e-5 - theta_t[i]) / direc;
  else
    alim = 1.0;
  if (alim < alim1)    // Ensure that all parameters are within the bounds
    alim1 = alim;
}
```

In the previous code, alim1 will be the intersection (minimum) of all possible 0-1 limitations and hence guarantees that all unique variance parameters are legal. If the direction for one of these parameters is positive, the upper limit of 1.0 will be our concern, so we keep it away from one by 1.e-5. If the direction is negative, hitting the lower bound of zero is the concern. Otherwise, we have no limit problem for this parameter. By keeping track of the minimum multiplier across all parameters, we guarantee that no parameter will go outside its legal bound.

The offset of 1.e-5 is not critical, except for one thing. The EMstep() code shown on page 236 forced the computed unique variances to be 1.e-6 away from the 0-1 bound. This QUADstep() code must keep it a bit further away. Otherwise, QUADstep() could set a point outside the EMstep() limit, and if this point happens to be the winner and hence be kept, then EMstep() might force a backtrack. This would complicate convergence tests. In fact, there is nothing wrong with QUADstep() forcing the point to be even further inside the limits, perhaps a lot further, because there is no danger in doing this. All we are doing here is defining the positions of the three new points that form the basis of the quadratic fit. There's not much critical about that, as long as the points are spaced far enough apart to ensure good numerical accuracy in computing the fit.

Now that we have a multiplier that is as close to the optimal 1.0 as possible, yet without violating any bounds, we can use Equation (4.23) on page 240 to compute the first of these three new trial points. The following steps are taken:

- If the step distance out from θ_t is so small that computation of the quadratic fit would be ill conditioned, we flag this fact so that we do not try the fit later. It would be reasonable to quit right here, instead of going on to the second point as I do in my implementation. However, continuing sometimes pays off, as the second or third point can often have superior log likelihood. Besides, the situation of a tiny multiplier is uncommon, so the issue is largely moot anyway.

- Evaluate the log likelihood (LL_1) at this first of the three new points. If it sets a new record, update the record and save these superior parameters in best_theta.

- In the extremely rare case (I've never seen it happen) that the log likelihood function has a catastrophic failure, set the ill_conditioned flag to prevent an attempt at a quadratic fit later.

```
if (alim1 < 0.01)  // Points must be far enough apart to get a good quadratic curve
   ill_conditioned = 1;
else {
   for (i=0; i<nparams; i++) {
      direc = theta_t[i] - theta_tm1[i];
      trial_theta[i] = theta_t[i] + alim 1 * direc;   // Equation (4.23)
   }
```

```
LL1 = log_lik_fast (trial_theta);
if (LL1 > *LLret) {
  *LLret = LL1;
  memcpy (best_theta, trial_theta, nparams * sizeof(double));
  }

if (LL1 < -1.e50)
  ill_conditioned = 1;
}
```

The other two new points are similarly constructed; this redundant code is omitted here but can be found in AN_FACTOR.TXT. Before continuing to the quadratic fit, we make sure that the ill_conditioned flag has not been set. If all is good, we compute the five quadratic fit coefficients using Equations (4.26) through (4.30), which start on page 240.

```
if (ill_conditioned)  // We need all six points to be good to proceed
  goto QUAD_FINISH;

aa = (LL1 - LL_t - alim1 * alim1 * (LL_tm1 - LL_t)) / (alim1 + alim1 * alim1);
bb = (LL2 - LL_t - alim2 * alim2 * (LL_tm2 - LL_t)) / (alim2 + alim2 * alim2);
cc = LL_tm1 - LL_t + aa;
dd = LL_tm2 - LL_t + bb;
ee = -0.5 * (LL3 - LL_t - (aa-bb) * alim3 - (cc + dd) * alim3 * alim3) / (alim3 * alim3);
```

Equation (4.31) on page 241 requires $\mathbf{H}^{-1}$, but we use the simple direct formula, because it is just two-by-two. We could even simplify the code a bit more by skipping the intermediate step of inverting the matrix, but it's clearer this way. The determinant of the matrix is an important indicator of the situation. In the extremely unlikely event that the determinant is positive, we have a minimum instead of a maximum, so don't bother continuing! If the determinant is tiny, the fit is too ill-conditioned to be worth pursuing.

```
// Invert two-by-two H matrix
det = cc * dd - ee * ee;
if (det > -1.e-12)
  goto QUAD_FINISH;

cci = dd / det;     // Upper-left diagonal of inverse
ddi = cc / det;     // Lower-right
eei = -ee / det;    // Off-diagonal
```

// Compute x and y, the max or saddle point of this quadratic fit, using Equation (4.31)
x = -0.5 * (aa * cci + bb * eei);
y = -0.5 * (aa * eei + bb * ddi);

Now we have to use the same procedure that we used for the three new points, expressing this stationary (and ideally maximum) as θ_t plus a multiplier times the direction of the stationary point. With any luck, the multiplier can be 1.0 so that we can evaluate the log likelihood at exactly the stationary point (and ideally maximum versus just saddle point) of this quadratic fit. But we may have to shrink the multiplier below one in order to avoid violating the 0-1 constraint on one or more unique variances. We saw this expressed in Equation (4.32) on page 241. The code to do this is shown next. It is similar to what we saw earlier for the three new points. Then we just retrieve the best parameters. We're done.

```
alim4 = 1.0;
for (i=0; i<npred; i++) {
  direc = x * (theta_t[i] - theta_tm1[i]) + y * (theta_t[i] - theta_tm2[i]);
  if (direc > 0.0)
    alim = (1.0 - 1.e-5 - theta_t[i]) / direc;
  else if (direc < 0.0)
    alim = (1.e-5 - theta_t[i]) / direc;
  else
    alim = 1.0;
  if (alim < alim4)
    alim4 = alim;
  }
if (alim4 < 0.01)     // Not worth another expensive log likelihood eval if this close
  goto QUAD_FINISH;
else {
  for (i=0; i<nparams; i++) {
    direc = x * (theta_t[i] - theta_tm1[i]) + y * (theta_t[i] - theta_tm2[i]);
    trial_theta[i] = theta_t[i] + alim 4 * direc;   // Equation (4.32)
    }
  LL4 = log_lik_fast (trial_theta);
  if (LL4 > *LLret) {
    *LLret = LL4;
    memcpy (best_theta, trial_theta, nparams * sizeof(double));
    }
  }
```

```
QUAD_FINISH:
  memcpy (PSIvec, best_theta, npred * sizeof(double));
  memcpy (Amat, best_theta+npred, npred * n_dim * sizeof(double));
}
```

Putting It All Together

In this section we'll present an overview, along with numerous code fragments, about how to assemble the routines just seen into a complete routine for performing my modified version of maximum likelihood factor analysis. The full code, except for error handling, can be found in AN_FACTOR.TXT. We begin with the class declaration:

```
class AnalyzeFactorChild {

public:
  AnalyzeFactorChild (int npreds, int *preds, int n_dim, int nonpar);
  ~AnalyzeFactorChild ();
  int AnalyzeFactorChild::EMstep ();
  void AnalyzeFactorChild::QUADstep (double *LL);
  double AnalyzeFactorChild::log_lik (double *theta);
  double AnalyzeFactorChild::log_lik_fast (double *theta);

  int error;              // Flags any error during constructor

  int npred;              // Number of predictors
  int n_dim;              // User-specified number of dimensions
  int preds[MAX_VARS];    // Database indices of predictors
  int nonpar;             // Use nonparametric correlation for tail control?

  // Work areas for optimization

  double *covar;          // Covariance (correlation) matrix
  double *Amat;
  double *Fmat;
  double *Gmat;
  double *Hmat;
  double *PSIvec;
  double *TEMPmat1;
  double *TEMPmat2;
```

```
    double *invert_rwork;
    int *invert_iwork;

    // Work areas specifically for QUADstep
    double *theta_t;
    double *theta_tm1;
    double *theta_tm2;
    double *trial_theta;
    double *best_theta;
    double LL_t;
    double LL_tm1;
    double LL_tm2;
    double LL1;
    double LL2;
    double LL3;
    double LL4;
};
```

There are a few global variables that hold information about this process and its results. The purpose of these variables is to facilitate subsequent operations such as rotation or display. They are declared external here.

```
extern int eigen_npred;     // Number of variables (generally predictors)
extern int eigen_preds[MAX_VARS]; // Their indices in database
extern int eigen_n_dim;     // User-specified number of unobserved factors
extern double *eigen_evals;
extern double *eigen_structure;
extern double *eigen_phi;
```

We make local and global copies of the calling parameters. The error flag will be set to a nonzero quantity if there is an error during the constructor call.

```
eigen_npred = npred = np;
eigen_n_dim = n_dim = nd;
nonpar = nonp;
for (i=0; i<np; i++)
  eigen_preds[i] = preds[i] = p[i];

error = 0;
```

Back when the EMstep() and QUADstep() routines were presented, they referenced numerous arrays that we had to trust were properly allocated. We now see these allocations. The global variables need to be freed (or just reallocated, if that's your preference) because their sizes may change now from what they were previously.

```
if (eigen_evals != NULL)
  FREE (eigen_evals);
if (eigen_structure != NULL)
  FREE (eigen_structure);
if (eigen_phi != NULL)
  FREE (eigen_phi);

val = (double *) MALLOC (npred * sizeof(double));
eigen_evals = (double *) MALLOC (npred * sizeof(double));
eigen_structure = (double *) MALLOC (npred * npred * sizeof(double));
eigen_phi = (double *) MALLOC (npred * sizeof(double));
work1 = (double *) MALLOC (npred * sizeof(double)); // For means and evec_rs()
work2 = (double *) MALLOC (npred * sizeof(double)); // For stddev
covar = (double *) MALLOC (npred * npred * sizeof(double));
Amat = (double *) MALLOC (npred * n_dim * sizeof(double));
Fmat = (double *) MALLOC (npred * n_dim * sizeof(double));
Gmat = (double *) MALLOC (npred * n_dim * sizeof(double));
Hmat = (double *) MALLOC (npred * n_dim * sizeof(double));
PSIvec = (double *) MALLOC (npred * sizeof(double));
TEMPmat1 = (double *) MALLOC (npred * npred * sizeof(double));
TEMPmat2 = (double *) MALLOC (npred * npred * sizeof(double));
invert_rwork = (double *) MALLOC ((npred * npred + 2 * npred) * sizeof(double));
invert_iwork = (int *) MALLOC (npred * sizeof(int));
k = npred * n_dim + npred; // Number of parameters (Psi plus A)
theta_t = (double *) MALLOC (5 * k * sizeof(double));
theta_tm1 = theta_t + k;
theta_tm2 = theta_tm1 + k;
trial_theta = theta_tm2 + k;
best_theta = trial_theta + k;
```

```
if (nonpar)
  nonpar_work = (double *) MALLOC (2 * n_cases * sizeof(double));
else
  nonpar_work = NULL;
```

If the user has requested that nonparametric correlation be used (to accommodate heavy-tailed data), we compute it here. See SPEARMAN.CPP for the computation routine.

```
if (nonpar) {
  k = 0;
  for (i=1; i<npred; i++) {
    for (j=0; j<i; j++) {
      for (icase=0; icase<n_cases; icase++) {
        nonpar_work[icase] = database[icase*n_vars+preds[i]];
        nonpar_work[n_cases+icase] = database[icase*n_vars+preds[j]];
        }
      covar[i*npred+j] = spearman (n_cases, nonpar_work, nonpar_work+n_cases,
                                   nonpar_work, nonpar_work+n_cases);
      ++k;
      }
    }
  }
```

Otherwise, we compute the mean and standard deviation and correlation matrix. It would be mathematically equivalent to directly compute the covariance matrix and then convert it to a correlation matrix, but that method has slightly less numerical stability. Note that although the correlation matrix is symmetric and evec_rs() ignores the redundant upper triangle, EMstep() is most efficient and clear when the entire matrix is filled in, so we copy the lower triangle to the upper.

```
else {
  for (i=0; i<npred; i++)
    work1[i] = work2[i] = 1.e-60;

  for (i=0; i<n_cases; i++) {
    for (j=0; j<npred; j++)
      work1[j] += database[i*n_vars+preds[j]];
    }
```

```
   for (j=0; j<npred; j++)
     work1[j] /= n_cases;    // Mean vector

   for (i=0; i<n_cases; i++) {
     for (j=0; j<npred; j++) {
       diff = database[i*n_vars+preds[j]] - work1[j];
       work2[j] += diff * diff;
       }
     }

   for (j=0; j<npred; j++)
     work2[j] = sqrt (work2[j] / n_cases);   // Standard deviation

// Compute correlation matrix 'covar'

   for (i=1; i<npred; i++) {
     for (j=0; j<i; j++)
       covar[i*npred+j] = 0.0;
     }

   for (i=0; i<n_cases; i++) {
     for (j=1; j<npred; j++) {
       diff = (database[i*n_vars+preds[j]] - work1[j]) / work2[j];
       for (k=0; k<j; k++) {
         diff2 = (database[i*n_vars+preds[k]] - work1[k]) / work2[k];
         covar[j*npred+k] += diff * diff2;
         }
       }
     }

   for (j=0; j<npred; j++) {
     for (k=0; k<j; k++)
       covar[j*npred+k] /= n_cases;
     }
   } // Else not nonpar, so compute means, stddev, correl

// The strict lower triangle has been computed. Fill in diagonal and upper triangle.
```

```
for (j=0; j<npred; j++) {
  covar[j*npred+j] = 1.0;
  for (k=j+1; k<npred; k++)
    covar[j*npred+k] = covar[k*npred+j]; // Needed for EMstep()
}
```

We now compute the eigenvalues and vectors of the correlation matrix and then compute the initial factor structure matrix by multiplying each eigenvector by the square root of its corresponding eigenvalue. We place all of them in the global area, although the first n_dim columns will be replaced with the factors later. Of more immediate importance is that we place the first n_dim columns in Amat, which will be the current estimate of the factor loadings throughout the algorithm.

```
evec_rs (covar, npred, 1, eigen_structure, eigen_evals, work1);

for (i=0; i<npred; i++) {
  for (j=0; j<npred; j++) {
    eigen_structure[i*npred+j] *= sqrt(eigen_evals[j]);
    if (eigen_structure[i*npred+j] < -1.0) // In a perfect fpt world would never happen
      eigen_structure[i*npred+j] = -1.0;
    if (eigen_structure[i*npred+j] > 1.0)
      eigen_structure[i*npred+j] = 1.0;
    if (j < n_dim)
      Amat[i*n_dim+j] = eigen_structure[i*npred+j];
  }
}
```

Compute the initial value of the Psi (Ψ) diagonal as was described on page 232. In particular, we implement Equation (4.15). Keep all of the unique variances away from zero, as many things become undefined or unstable at or near zero. We save these values in the global area, even though they will be overwritten later. It's silly, perhaps, but clean and clear. More importantly, we save them in PSIvec, which will hold the current values during optimization.

```
for (i=0; i<npred; i++) {
  eigen_phi[i] = 1.0;
  for (j=0; j<n_dim; j++)
    eigen_phi[i] -= eigen_structure[i*npred+j] * eigen_structure[i*npred+j];
  if (eigen_phi[i] < 1.e-3)
    eigen_phi[i] = 1.e-3;
  PSIvec[i] = eigen_phi[i]; // Initialize for optimization
  }
```

We come now to the heart of the matter, the iterative alternation of EMstep() and QUADstep(). When we get to QUADstep(), we'll need the log likelihood at three points: current (t), lag 1 ($tm1$), and lag2 ($tm2$). These are as follows:

```
theta_t       LL_t
theta_tm1     LL_tm1
theta_tm2     LL_tm2
```

So we initialize by letting the starting values just computed be the oldest point, and then we run one EMstep() to be the second oldest. When we get inside the loop, we'll begin the loop with an EMstep(), which will give the current point. Here is the initialization code. Note that the values computed now will be shifted back one time slot inside the loop. Also recall that PSIvec and Amat are the current values of the parameters as optimization progresses, and they serve as both input to and output from EMstep().

```
memcpy (theta_tm1, PSIvec, npred * sizeof(double));
memcpy (theta_tm1+npred, Amat, npred * n_dim * sizeof(double));
LL_tm1 = log_lik_fast (theta_tm1);

if (EMstep ()) {
  // Issue error message here; this error is extremely unlikely
  goto FACTOR_FINISH;
  }

memcpy (theta_t, PSIvec, npred * sizeof(double));
memcpy (theta_t+npred, Amat, npred * n_dim * sizeof(double));
LL_t = log_lik_fast (theta_t);

EMreverse = 0;  // Will count rare pathological event that can cause endless looping
```

Preparation for the iteration is complete. We have the log likelihood computed at two points and stored in the current (t) and lag 1 ($tm1$) slots. For cleanliness, we place a limit on looping. In practice, we will never come even close to this limit. The optimization loop now begins.

The first step in the loop is to perform an EMstep(), which modifies the current values of PSIvec and Amat to be an improvement. Then we shift the two most recent points (t and $tm1$) and their log likelihoods back one time slot into the past and update the current point.

```
for (iter=0; iter<10000; iter++) {

  if (EMstep ()) { // This takes and returns PSIvec and Amat without touching theta_t
    // Issue error message here
    break;
  }
  memcpy (theta_tm2, theta_tm1, npred * sizeof(double));
  memcpy (theta_tm2+npred, theta_tm1+npred, npred * n_dim * sizeof(double));
  LL_tm2 = LL_tm1;

  memcpy (theta_tm1, theta_t, npred * sizeof(double));
  memcpy (theta_tm1+npred, theta_t+npred, npred * n_dim * sizeof(double));
  LL_tm1 = LL_t;

  memcpy (theta_t, PSIvec, npred * sizeof(double)); // EMstep() computed this
  memcpy (theta_t+npred, Amat, npred * n_dim * sizeof(double));
  LL_t = log_lik_fast (theta_t);
```

We check here for an unusual but possible pathological situation. If one or more of the unique variances (PSIvec) are extremely close to their 0-1 bound and EMstep() wants to drive them even closer, past the threshold built into the algorithm, then the value may bounce back and forth endlessly, pushed past the threshold by the EM algorithm and then snapped back by my modification that keeps them all away from the boundary. Count occurrences of this and abort if necessary.

```
  if (LL_t < LL_tm1) {
    ++EMreverse;
    if (EMreverse > 10) {
      // Issue error message here
      break;
    }
  }
```

At this point we have our three points, so we can call QUADstep(). Then we shift the former current value back one time slot and update the current value. There is no need to copy *tm1* to *tm2* as we did after EMstep() because EMstep() does not need any lagged values.

```
QUADstep (&LL); // Takes t, tm1, and tm2 as input and computes PSIvec, Amat

memcpy (theta_tm1, theta_t, npred * sizeof(double));
memcpy (theta_tm1+npred, theta_t+npred, npred * n_dim * sizeof(double));
LL_tm1 = LL_t;  // This came from EM above
memcpy (theta_t, PSIvec, npred * sizeof(double));
memcpy (theta_t+npred, Amat, npred * n_dim * sizeof(double));
LL_t = LL;  // This came from the QUADstep we just did
```

At this point, *tm1* is after the most recent EMstep(), *t* is after this QUADstep(), and *tm2* is still after the EMstep() before the most recent EMstep().

The final step in the loop is to check for convergence. It is dangerous to use changes in the log likelihood as a convergence test (though many do) because this function can become extremely flat near the optimum. So instead we base the test on the maximum change in any parameter after a set of three optimization steps, QUADstep(), EMstep(), and QUADstep(). (It really is three instead of what appears at first glance to be two; walk through the code if you don't believe me.)

```
max_change = 0.0;
for (i=0; i<npred+npred*n_dim; i++) {
  diff = fabs (theta_t[i] - theta_tm2[i]);
  if (diff > max_change)
    max_change = diff;
  }
if (max_change < 1.e-6)        // Fairly arbitrary choice
  ++convergence_counter;
else
  convergence_counter = 0;
if (convergence_counter > 2)   // Fairly arbitrary choice
  break;
}
```

After convergence is obtained, we copy the class variables containing the unique variances and factor loadings to the global area. Compute eigen_evals as the squared length of each column; it's not really an eigenvalue, but the resemblance is there, and we'll make some use of this in a moment.

```
for (i=0; i<npred; i++) {
  eigen_phi[i] = PSIvec[i];
  for (j=0; j<n_dim; j++)
    eigen_structure[i*npred+j] = Amat[i*n_dim+j];
}

for (j=0; j<n_dim; j++) {
  sum = 0.0;
  for (i=0; i<npred; i++)
    sum += Amat[i*n_dim+j] * Amat[i*n_dim+j];
  eigen_evals[j] = sum;
}
```

Sometimes it can be useful to see the factor loadings with the columns sorted from most to least prominent, as is the case for raw principal components. Note that this is not as useful as may seem, because unlike principal components, *factor loadings are not unique* and do not necessarily come out of the optimization algorithm in any particular order. Because we do initialize the loading to be principal components, there is usually a strong resemblance. But the factor loadings are unique only up to rotation; they define a unique subspace, but orthogonal rotations within that subspace give identical values for the log likelihood. So if you are interested in the loadings, it often pays to do a rotation such as varimax after computing them.

The code on the next page is a crude but simple algorithm for sorting the columns according to their squared length. Last but not least, we free all of the work areas.

```
for (i=1; i<n_dim; i++) {
  im1 = i - 1;
  ibig = im1;
  big = eigen_evals[im1];
  /* Find largest eval beyond im1 */
```

```
    for (j=i; j<n_dim; j++) {
      if (eigen_evals[j] > big) {
        big = eigen_evals[j];
        ibig = j;
        }
      }

    if (ibig != im1) {  // Do we need to swap ibig and im1?
      eigen_evals[ibig] = eigen_evals[im1];
      eigen_evals[im1] = big;

      for (j=0; j<npred; j++) {
        sum = eigen_structure[j*npred+im1];
        eigen_structure[j*npred+im1] = eigen_structure[j*npred+ibig];
        eigen_structure[j*npred+ibig] = sum;
        }
      }
    }

FACTOR_FINISH:
  FREE (covar);
  FREE (work1);
  FREE (work2);
  FREE (Amat);
  FREE (Fmat);
  FREE (Gmat);
  FREE (Hmat);
  FREE (PSIvec);
  FREE (TEMPmat1);
  FREE (TEMPmat2);
  FREE (invert_rwork);
  FREE (invert_iwork);
  FREE (theta_t);
  if (nonpar_work != NULL)
    FREE (nonpar_work);
}
```

Thoughts on My Version of the Algorithm

I've mentioned several times during this development that my version of the maximum-likelihood factor analysis algorithm is slightly different from the usual version, though not much, and easily revised to the standard version. The reason is that in my own work, *I am not so much interested in the factor loadings as the unique variances*. This lets me identify any variables that are members of highly redundant sets. Such variables can be removed or given special treatment. One particularly useful approach is to collect all variables with unique variance near zero and compute their most dominant principal components. This provides a few very nonredundant variables to replace many redundant variables, usually with negligible loss of information.

Since a measure of uniqueness versus redundancy is my primary goal, I am motivated to standardize the variables before beginning the factor analysis and then enforce a rigid 0–1 constraint on the unique variances. This makes the computed values easy to interpret. The more usual approach is to ensure that the variables are roughly commensurate before conducting the analysis, avoid standardization, and impose no upper limit on the unique variance.

If you want to implement the usual algorithm rather than mine, the changes in the code are almost trivial to implement. Skip standardization, computing the covariance matrix instead of the correlation matrix. In the code that computes the initial estimate of Psi, Equation (4.15) on page 233 will have to be evaluated with the actual diagonal of S, the variances, rather than 1.0, which is the diagonal of a correlation matrix. In the EMstep() code, remove the imposition of an upper bound of one. In the QUADstep() code do the same. That's it. But please understand that in the absence of standardization, convergence can be significantly slower than with standardized variables.

Measuring Coherence

It is often the case that a set of variables that are measured across time will have varying interrelationships. It may be that under "normal" circumstances they move in predictable patterns relative to one another. One example comes from the commodity futures markets. Long-range (several months ahead) weather predictions impact futures prices for grains, which in turn impact futures prices for meat products. If a time comes along in which their interrelationship falters, this is an indication that something funny is going on, and maybe we had better sit up and pay attention. In particular, if we are using a trained model to make predictions, we should consider whether this model is still valid.

The opposite situation can happen as well: time-series variables that normally have a certain degree of independence may suddenly begin to track unnaturally. The classic example of this is in the stock market. Frightening world events, such as talk of immanent war, may cause the prices of all market sectors to trend lower simultaneously, when under normal circumstances they tend to move somewhat independently.

Of course, these phenomena are not limited to financial applications. Suppose an assembly line monitors various recent (across a *lookback* window of time) parameters such as flow rate of various ingredients, temperature of heating chambers, color of final product as it rolls off the line, and so forth. Normally, these variables should have a fairly constant interrelationship. If we suddenly see this relationship disappear, we had better run some diagnostics on the line and see what's going on.

It should come as no surprise that there is an infinite number of ways to measure *coherence*, the degree to which a set of time-series variables are interrelated within a lookback window that moves forward as time progresses. One reasonable way is to determine how much of the standardized total variance is concentrated in the largest eigenvalue. (We should always standardize the variables so that individual offsets and scales do not impact our measurement.) The disadvantage of this approach is that it measures the degree to which coherent variation exists in a *single* dimension. Sometimes this is appropriate, so we should consider the largest eigenvalue as a possible measure of coherence. But in many or most applications, coherence may be represented by relationships in several dimensions. As a trivial example, we may have four variables, and their normal relationship may be that X_1 and X_2 are correlated, as are X_3 and X_4, while variables in the first pair have little or no relationship with those in the second pair. Examining just the largest eigenvalue will miss this dual relationship since a single eigenvector cannot represent both relationships.

This problem can be alleviated by considering the fraction of the total variance contained in the few largest eigenvalues. But this requires an assumption of how many relationships exist (the dimensionality of the relationship space). In many cases, one can do an eigenstructure analysis in advance, under normal conditions, and choose to use the number of dominant eigenvalues. This is a good approach when it is feasible.

I now present a more general approach that is appropriate when one does not have prior information concerning the number of valid relationships or when the number of relationships varies across time, a common occurrence when there is a large number of variables. This would be the case, for example, when we are studying the price changes of a large basket (a hundred or more) of equities. This method is superior under such conditions but inferior when the dimensionality is constant and we know what it is.

So if we happen to have a known fixed dimensionality, the best approach is to add that number of largest eigenvalues and divide by the sum of all eigenvalues (which will equal the number of variables if the variables are standardized).

A good way to approach the more general situation (no assumption of dimensionality) is to visualize the eigenvalues, sorted from largest to smallest, as sitting on a teeter-totter or balance-beam scale. Imagine that the largest eigenvalue is on the far left, the smallest on the far right, and the intermediates equally spaced in between. The coherence is the rotational force exerted on the beam caused by imbalance in the eigenvalues. We can compute this force as a weighted sum of the eigenvalues, with the weights defined by the equally spaced locations on the beam. The weights to the left of the center are positive, and the weights to the right of center are symmetrically negative.

Let's consider the two most extreme possibilities. Suppose every variable is completely independent of every other variable within our lookback window. Their correlation matrix will be an identity matrix, and the eigenvalues will all be equal (1.0). Because the weights given to each eigenvalue are symmetric around the center (in accord with the balance beam analogy), the weighted sum will be zero. Thus, the coherence in this totally uncorrelated situation will be zero. Note that a coherence less than zero is not possible, because the eigenvalues are sorted, with the larger values on the left (positive weights) side.

For convenience, we scale the weights such that the leftmost weight (that for the largest eignvalue) is 1.0, and that for the rightmost (the smallest eigenvalue) is -1.0. Now suppose the measured variables are all perfectly correlated with one another; they are all (possibly different) linear transformations of some underlying variable. There will be only one nonzero eigenvalue in this one-dimensional situation, and it will equal the number of variables. Hence, the weighted sum will be the number of variables (the leftmost weight times this largest eigenvalue). If we normalize the weighted sum by dividing it by the number of variables, we see that the coherence in this situation of all variables being perfectly correlated with one another is 1.0.

Thus, we have a 0-1 measure of the degree to which a set of variables have correlations among themselves, as defined by the imbalance in their eigenvalue distribution. This measure makes no assumptions on the dimensionality of the underlying structure.

Note that in real life, random variation will cause variables that are truly uncorrelated to have some measured correlation, especially if the lookback window is short. Any correlation at all among the measured variables will cause some imbalance in the eigenvalues; the only way they can all be equal (and hence achieve perfect balance) is if all off-diagonal correlations are exactly zero. So in practice, the computed coherence has an unavoidable upward bias. But usually we are not interested in the actual coherence. In a data mining situation we are most concerned with stability across time: is the coherence reasonably constant? It is a sudden unexplained *change* in the coherence that merits our attention. That's the flag for employing multiple models or other remedial action.

Code for Tracking Coherence

We show here the essential code for computing coherence across a moving window. As usual, mundane things like error checking are omitted for clarity. The complete code can be found in the file AN_COHERENCE.CPP.

We begin with allocation of memory. The array val will hold the computed coherence values. All other allocations are temporary work areas. There are n_cases in the database, each consisting of a row of n_vars variables, from which we will select npred of them, indexed in preds. The moving window consists of lookback observations.

```
int icase, i, j, k;
double *dptr, *means, *evals, *evects, *workv, minval, maxval, meanval;
double sum, total, diff, diff2, *nonpar_work, factor;
char msg[512], line[1024], coherence_log[1024];
FILE *fp;

val = (double *) MALLOC ((n_cases-lookback+1) * sizeof(double));
means = (double *) MALLOC (npred * sizeof(double));
covar = (double *) MALLOC (npred * npred * sizeof(double));
evals = (double *) MALLOC (npred * sizeof(double));
evects = (double *) MALLOC (npred * npred * sizeof(double));
workv = (double *) MALLOC (npred * sizeof(double));
if (nonpar)  // Did the user request nonparametric correlation?
  nonpar_work = (double *) MALLOC (2 * lookback * sizeof(double));
else
  nonpar_work = NULL;

/*
  Get ready to write coherence values to a file
*/
```

```
_fullpath (coherence_log, "COHERENCE.TXT", 1024); // Will write coherences here
if (fopen_s (&fp, coherence_log, "wt")) {
   // Handle error messages here
   goto COHERENCE_FINISH;
   }
```

This is the main loop that processes all cases. We'll keep track of the minimum, maximum, and mean coherences to report to the user.

```
/*
   Main outer loop does all cases
*/

   minval = 1.e30;
   maxval = -1.e30;
   meanval = 0.0;

   for (icase=lookback-1; icase<n_cases; icase++) {
```

If the user requested nonparametric correlation, compute it here. We need only the lower minor triangle of the symmetric correlation matrix.

```
   if (nonpar) {
      covar[0] = 1.0;                      // First diagonal entry
      for (i=1; i<npred; i++) {
         for (j=0; j<i; j++) {             // Just do lower minor triangle
            for (k=0; k<lookback; k++) {   // Traverse the moving window
               dptr = database + n_vars * (icase - k);    // Point to this case in database
               nonpar_work[k] = dptr[preds[i]];           // Get one variable
               nonpar_work[lookback+k] = dptr[preds[j]];  // And the other
               }
            covar[i*npred+j] = spearman (lookback, nonpar_work,  // In SPEARMAN.CPP
                     nonpar_work+lookback, nonpar_work, nonpar_work+lookback);
            }
         covar[i*npred+i] = 1.0;           // Diagonal of a correlation matrix is 1.0
         }
      }
```

If the user did not request nonparametric correlation, compute the covariance matrix and then convert it to a correlation matrix. First we must compute the means to center the data.

```
else {
  for (i=0; i<npred; i++)
    means[i] = 0.0;

  for (i=0; i<lookback; i++) {                  // Compute means across window
    dptr = database + n_vars * (icase - i);     // Point to this case in database
    for (j=0; j<npred; j++)
      means[j] += dptr[preds[j]];
    }

  for (j=0; j<npred; j++)
    means[j] /= lookback;
```

Now compute the covariance matrix and convert it to a correlation matrix.

```
  for (i=0; i<npred; i++) {
    for (j=0; j<=i; j++)
      covar[i*npred+j] = 0.0;
    }

  for (i=0; i<lookback; i++) {
    dptr = database + n_vars * (icase - i);      // Point to this case in database
    for (j=0; j<npred; j++) {                     // One variable
      diff = dptr[preds[j]] - means[j];           // Center it
      for (k=0; k<=j; k++) {                       // Lower triangle, including diagonal
        diff2 = dptr[preds[k]] - means[k];         // Center the other variable
        covar[j*npred+k] += diff * diff2;          // Definition of covariance
        }
      }
    }

  for (j=0; j<npred; j++) {
    for (k=0; k<=j; k++)
      covar[j*npred+k] /= lookback;
    }
```

```
for (j=1; j<npred; j++) {          // Convert lower minor triangle to correlations
  for (k=0; k<j; k++)
    covar[j*npred+k] /= sqrt (covar[j*npred+j] * covar[k*npred+k]);
  }

for (j=0; j<npred; j++)            // Diagonal is unity
  covar[j*npred+j] = 1.0;

} // Else not nonpar, so compute means and covar, correlation
```

Compute the eigenvalues of the correlation matrix. Compute the coherence and store it in val for display and writing to a file. The total is the sum of all eigenvalues, which theoretically equals npred, so this is a minor waste but helps with clarity and tiny floating-point errors.

```
evec_rs (covar, npred, 0, evects, evals, workv);    // In EVEC_RS.CPP

factor = 0.5 * (npred - 1);             // Center of balance beam
sum = total = 0.0;
for (i=0; i<npred; i++) {
  total += evals[i];                    // Not really needed
  sum += (factor - i) * evals[i] / factor; // Coherence is weighted sum
  }

// Compute and save the criterion
sum /= total;
val[icase-lookback+1] = sum;

if (val[icase-lookback+1] > maxval)
  maxval = val[icase-lookback+1];
if (val[icase-lookback+1] < minval)
  minval = val[icase-lookback+1];
meanval += val[icase-lookback+1];

} // For all cases
```

Coherence in the Stock Market

On the next page I show coherence plots for just three stocks, BAC, DOW, and IBM, which represent very different market sectors. Both use nonparametric correlation of daily market changes. The top plot has a lookback of 50 days, and the bottom 252 days (about one year of trading).

   One thing that pops out is the tremendous range of coherence. With just 50 days, the coherence ranges from practically zero to almost 0.9, and even with a year of lookback it still varies tremendously. The sudden sharp spike just before case 1000 is Black Monday (October 19, 1987). Surely there is useful information to data mine here!

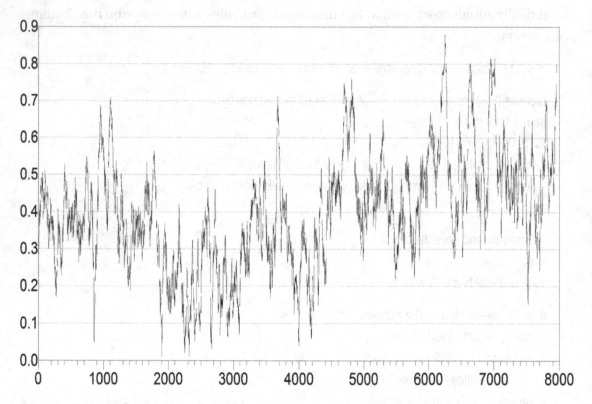

Figure 4-2. *Coherence with lookback=50*

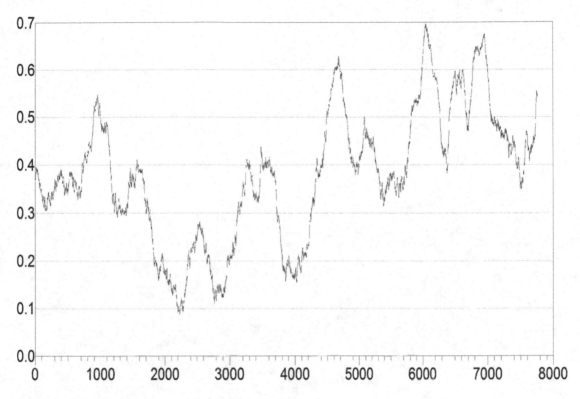

Figure 4-3. *Coherence with lookback=252*

CHAPTER 5

Using the DATAMINE Program

This chapter serves as a user's manual for the DATAMINE program, which demonstrates the algorithms presented in this book. Each menu selection is discussed in its own section.

File/Read Data File

A text file in standard database format is read. In particular, standard-format Excel CSV files may be read, as well as databases produced by many common statistical and data analysis programs. The first line must specify the names of the variables in the database. The maximum length of each variable name is 15 characters. The name must start with a letter and may contain only letters, numbers, and the underscore (_) character.

Subsequent lines contain the data, one case per line. Missing data is not allowed.

Spaces, tabs, and commas may be used as delimiters for the first (variable names) and subsequent (data) lines.

Here are the first few lines from a typical data file. Six variables are present, and three cases are shown.

```
RAND0 RAND1 RAND2 RAND3 RAND4 RAND5
-0.82449359   0.25341070   0.30325535  -0.40908301  -0.10667177   0.73517430
-0.47731471  -0.13823473  -0.03947150   0.34984449   0.31303233   0.66533709
 0.12963752  -0.42903802   0.71724504   0.97796118  -0.23133837   0.81885117
```

© Timothy Masters 2018
T. Masters, *Data Mining Algorithms in C++*, https://doi.org/10.1007/978-1-4842-3315-3_5

File/Exit

The program is terminated.

Screen/Univariate Screen

The algorithm described starting on page 110 is used to screen a set of predictor candidates for a relationship with a single target. The menu shown in Figure 5-1 will appear.

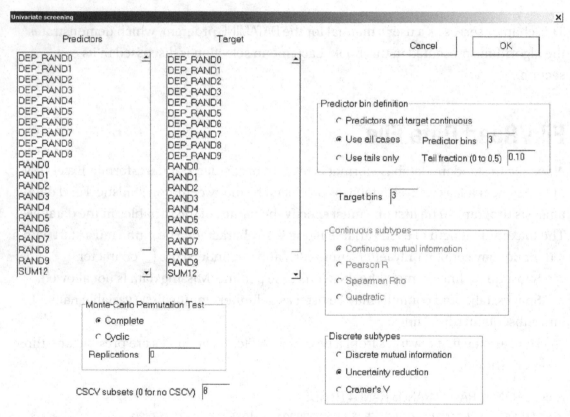

Figure 5-1. *Univariate screening*

The user must make the following selections and specifications:

- *Predictors*: Select a set of predictor candidates to be tested for a relationship with a single target.

- *Target*: Select a single target.

- *Predictor bin definition*: Specify the nature of the predictors (and by extension, the target). The choices are as follows:

  - *Predictors and target continuous*: All variables are to be treated as continuous.

  - *Use all cases*: All variables are treated as discrete. Continuous variables are converted to discrete bins. The user must specify the number of bins to use for the predictors.

  - *Use tails only*: The predictors are split into two bins: the tails (extreme values). The user must specify the fraction of extreme values to keep in each tail.

- *Target bins*: If the user selected either of the discrete options (*Use all cases* or *Use tails only*), then this specifies the number of bins into which the target variable is categorized.

- *Continuous subtypes*: If the user selected *Predictors and target continuous*, you specify the relationship criterion to be used. See the section beginning on page 77.

- *Discrete subtypes*: If the user selected either of the discrete options above (*Use all cases* or *Use tails only*), then this specifies the relationship criterion to be used. See the section beginning on page 77.

- *Monte Carlo Permutation Test*: A *Replications* value greater than 1 will cause a Monte Carlo permutation test to be performed, with this many tests run, one of which is unpermuted. The user also specifies the type of permutation, *Complete* or *Cyclic*. This topic is discussed starting on page 89.

- *CSCV subsets*: This controls performance of the CSCV test, discussed starting on page 97.

Screen/Bivariate Screen

This section discusses bivariate screening, in which we search for relationships between one or more predictor candidates and one or more target candidates. The menu shown in Figure 5-2 will appear.

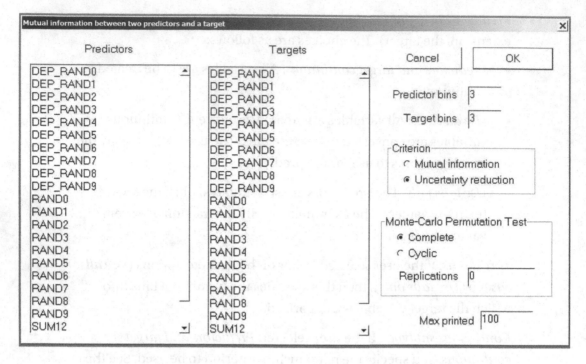

Figure 5-2. *Bivariate screening*

The user must make the following selections and specifications:

- *Predictors*: Select a set of predictor candidates to be tested for pairwise relationships with one or more targets.

- *Target*: Select a set of targets to be tested for a relationship with pairs of predictors.

- *Predictor bins*: This specifies the number of bins into which the predictor variables are categorized.

- *Target bins*: This specifies the number of bins into which the target variables are categorized.

- *Criterion*: The user chooses whether the relationship criterion is mutual information (page 17) or uncertainty reduction (page 61).

- *Monte Carlo Permutation Test*: A *Replications* value greater than 1 will cause a Monte Carlo permutation test to be performed, with this many tests run, one of which is unpermuted. The user also specifies the type of permutation, *Complete* or *Cyclic*. This topic is discussed starting on page 89.

- *Max printed*: If the user specifies numerous predictors and targets, the number of combinations of pairs of predictors with individual targets can be enormous. A line in the DATAMINE.LOG file is printed for each such combination, sorted from best to worst. This option lets the user limit the number of lines printed, beginning with the best.

Screen/Relevance Minus Redundancy

This section discusses relevance-minus-redundancy screening, in which we use a forward stepwise search for relationships between a set of predictor candidates and a single target variable. This algorithm was discussed on page 124. The menu shown in Figure 5-3 will appear.

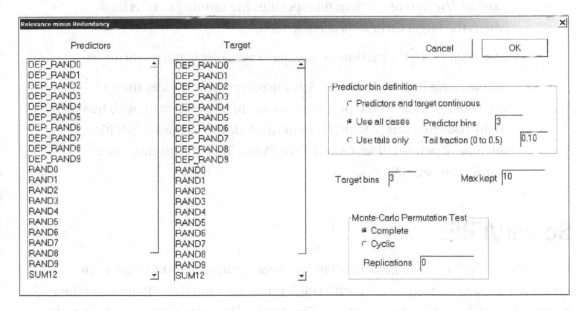

Figure 5-3. *Relevance-minus-redundancy screening*

The user must make the following selections and specifications:

- *Predictors*: Select a set of predictor candidates to be stepwise tested for inclusion in the set of predictors having maximum relationship with the target.

- *Target*: Select a single target to be tested for a relationship with a set of predictors.

- *Predictor bin definition*: Specify the nature of the predictors (and, by extension, the target). The choices are as follows:

  - *Predictors and target continuous*: All variables are to be treated as continuous.

  - *Use all cases*: All variables are treated as discrete. Continuous variables are converted to discrete bins. The user must specify the number of bins to use for the predictors.

  - *Use tails only*: The predictors are split into two bins: the tails (extreme values). The user must specify the fraction of extreme values to keep in each tail.

- *Target bins*: If the user selected either of the discrete options (*Use all cases* or *Use tails only*), then this specifies the number of bins into which the target variable is categorized.

- *Max kept*: This is the maximum number of variables in the predictor set.

- *Monte Carlo Permutation Test*: A *Replications* value greater than 1 will cause a Monte Carlo permutation test to be performed, with this many tests run, one of which is unpermuted. The user also specifies the type of permutation, *Complete* or *Cyclic*. This topic is discussed starting on page 89.

Screen/FREL

The *Feature Weighting as Regularized Energy-Based Learning* (FREL) algorithm presented starting on page 141 is used to rank predictor candidates in terms of their relationship with a single target variable. This method is particularly useful when the data is fairly clean (noise-free) but has relatively few cases compared to the number of predictor candidates. The menu screen shown in Figure 5-4 appears.

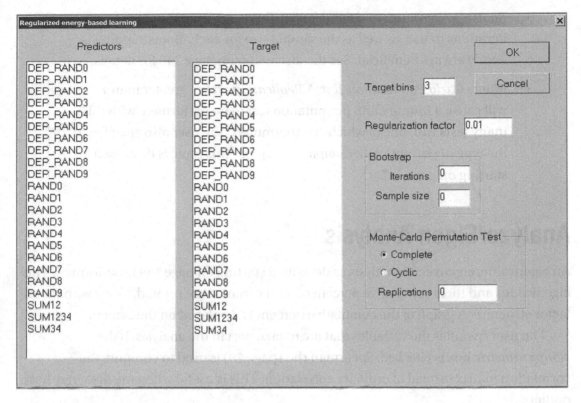

Figure 5-4. *FREL screening*

The user must make the following selections and specifications:

- *Predictors*: Select a set of predictor candidates to be ranked in terms of their relationship with the target.

- *Target*: Select a single target to be tested for a relationship with a set of predictors.

- *Target bins*: This specifies the number of bins into which the target variable is categorized.

- *Regularization factor*: This controls penalization for excessively large weights in the ranking scores. It is legal and computationally harmless to set this to zero. A general discussion of this parameter appears on page 145. Also see a more specific example of its use on page 151.

273

- *Bootstrap iterations and Sample size*: This is the number of bootstrap iterations to use, as well as the sample size for each. Bootstrapping is nearly always beneficial. See the discussion on page 146 for details.

- *Monte Carlo Permutation Test*: A *Replications* value greater than 1 will cause a Monte Carlo permutation test to be performed, with this many tests run, one of which is unpermuted. The user also specifies the type of permutation, *Complete* or *Cyclic*. This topic is discussed starting on page 147.

Analyze/Eigen Analysis

An eigenvalue/eigenvector analysis as described starting on page 189 is performed. The eigenvalues and their cumulative percent of total variance are printed, along with the factor structure. A graph of the cumulative percent is displayed on the screen.

The user specifies the variables that are to take part in the analysis. If the *Nonparametric* box is checked, Spearman rho (page 79) is used to compute the correlation matrix instead of ordinary correlation. This is useful when the data may have outliers.

Analyze/Factor Analysis

A maximum-likelihood factor analysis as described starting on page 221 is performed. The eigenvalues and their cumulative percent of total variance are printed first, along with the factor structure and initial Psi estimates (basic communalities). A graph of the cumulative percent is displayed on the screen. Then, the final factor analysis information is printed. Note that the *Squared length* printed at the top of each column of factor loadings is roughly analogous to the eigenvalues for an ordinary principal components analysis, but only roughly. This is because these factors are unique only up to rotation, so the natural ordering seen with the eigenvalues is no longer guaranteed.

The user specifies the variables that are to take part in the analysis. If the *Nonparametric* box is checked, Spearman rho (page 79) is used to compute the correlation matrix instead of ordinary correlation. This is useful when the data may have outliers.

Analyze/Rotate

If the user has performed either an *Eigen analysis* or a *Factor analysis*, a varimax factor rotation (page 199) may be performed. The menu shown in Figure 5-5 appears.

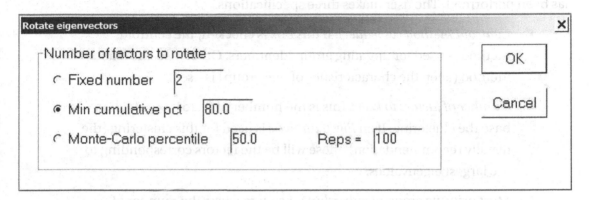

Figure 5-5. *Rotate eigenvectors*

The user must specify the number of factors to rotate. If the starting factors are from an *Eigen analysis*, we rotate the factor loadings corresponding to the specified number of largest eigenvalues. If they are from a *Factor analysis*, fully sensible results are obtained only if the user specifies the fixed number of factors that were computed in the factor analysis.

There are three ways to specify the number of factors to be rotated:

- A fixed number

- Those (starting from the largest eigenvalue) that make up the specified minimum percent of total variance.

- Horn's algorithm, described on page 202, determines the number of factors to keep. In this case, the percentile and number of replications must be specified.

Analyze/Cluster Variables

The technique described starting on page 213 is used to cluster variables. This operation may be invoked only if an *Eigen analysis* (most sensible) or *Factor analysis* (less sensible) has been performed. The user makes three specifications.

- *Centroid method (vs leader)*: If this box is checked, the centroid method is used for updating group identifiers. Otherwise, the leader method (keep the characteristics of one group) is used.

- *Number of factors to keep*: This is the number of factors on which to base the clustering. If an *Eigen analysis* is used for this clustering (the usually recommendation), these will be the factors corresponding to the largest eigenvalues.

- *Start printing group membership when n reaches*: The number of groups starts out at the number of variables. Each time a group is absorbed, the program can print group membership information. Obviously, this can result in a huge printout if the number of variables is large. This option lets the user specify that group membership printing does not begin until this many groups remain.

Analyze/Coherence

A time-domain coherence analysis, as described on page 257, is performed. The user specifies the variables that are to take part (which must be aligned in time) as well as the following parameters:

- *Connect*: If this box is checked, the plotted coherence values are connected. Otherwise, they are discrete vertical bars.

- *Nonparametric*: If this box is checked, Spearman rho (page 79) is used to compute the correlation matrix. Otherwise, it is computed with ordinary correlation. This option is recommended if the data may have outliers.

- *Lookback window cases*: This many of the most recent cases are used in the moving window for computation of coherence within the window. Longer windows result in more accurate measurements but poorer location in time.

Plot/Series

This just plots a time series of a single variable selected by the user. If the *Connected* box is checked, the plotted points are connected. Otherwise, each point is represented by a discrete vertical line.

Plot/Histogram

This plots a histogram of a single variable selected by the user. The user may optionally request that the lower and/or upper bounds of the plot be limited to specified values. If this is not done, the actual plot limits are at or slightly outside the full range of the variable. The user also specifies the number of bins to use.

Plot/Density

A plot for revealing relationship anomalies, as discussed starting on page 167, is done. The menu shown in Figure 5-6 appears.

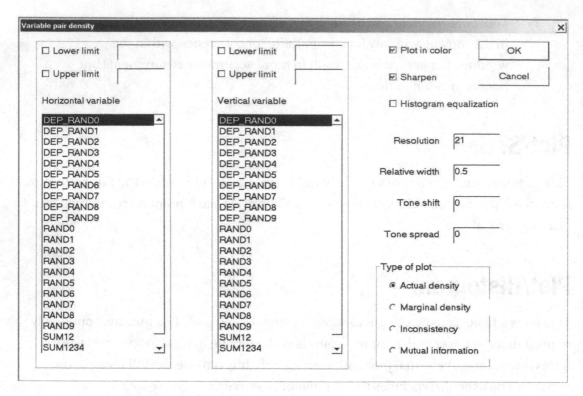

Figure 5-6. *Variable pair density*

The user specifies the following items:

- *Horizontal variable*: This is the variable that will be represented by the horizontal axis. The user may optionally check the *Lower limit* and/or the *Upper limit* box above this list and specify a numeric value (values) for display limits. If a box is not checked, the corresponding limit is at or slightly outside the actual range of the variable.

- *Vertical variable*: This specifies the variable for the vertical axis, as described.

- *Plot in color*: If this box is selected, the plot will be in color, with yellow indicating large values of the plotted quantity and blue indicating small values. Otherwise, it is black-and-white, with black indicating large values and white indicating small values.

- *Sharpen*: If this box is selected, areas of unusually large concentration are made to stand out from the background by accentuating them at the expense of contrast in other areas.

- *Histogram equalization*: If this box is selected, the program applies a nonlinear transform to the data in such a way that every possible displayed tone or color occurs in the display in approximately equal quantity. The effect of this transformation is usually that small changes in the data are made more visible, while simultaneously reducing the prominence of large changes.

- *Resolution*: This is the number of horizontal and vertical divisions at which the plot is computed. Computation time is roughly proportional to the square of this value. Larger values can reveal more detail about the relationship between the variables.

- *Relative width*: This is the width of the Parzen smoothing window, relative to the standard deviation of each variable. Smaller values reveal more information but can also accentuate noise. If the data is noisy, large width values are appropriate to smooth out the noise.

- *Tone shift*: This moves the overall display range. A positive value shifts the tones in the "high" direction, and negative shifts tones toward the "low" direction. The default of zero produces no change.

- *Tone spread*: This expands or compresses the range of the display. The default of zero produces no change. Negative values are legal but rarely useful, as this compresses variation into a narrow range, making discrimination difficult. Positive values, rarely beyond five or so, expand the center of the display range while squashing the extremes. This emphasizes features in the interior of the grid range, at the expense of the extremes.

- *Actual density*: This plots the actual density, as discussed on page 171.

- *Marginal density*: This plots the marginal density product, as discussed on page 171.

- *Inconsistency*: This plots the marginal inconsistency, as discussed on page 171.

- *Mutual information*: This plots the contribution of each region to the total mutual information, as discussed on page 172.

Index

A

Adaptive partitioning
    actual counts, compute, 57
    algorithm coding, 50–51
    bin counts, 49
    bivariate density, 46
    bivariate distribution, 47, 50
    chi-square test, 49
        statistic, 49
        two-by-two, 49, 56, 60
    continuous data, 56
    continuous variables, 45
    currentDataStart, 58
    currentDataStop, 58
    discrete formula, 45
    four-by-four chi-square tests, 60
    indices, 51
    indices array, 53
    method, 42
    MUTINF_C.CPP, 51
    naive algorithms, 46
    nonrandom distribution, 49
    nonuniform data distribution, 56
    partitioning diagram, 47–48
    random variation, 46
    rearranging indices, 58
    rectangle off the stack, 53
    splitting across tied data, 50
    splitting tied cases, 52
    stack entries, 52, 53
    starting and stopping indices, 54
    subrectangle cases, 58–59
    TEST_DIS program, 46
    tunable parameters, 45
    two-by-two grid, 46
    two-by-two split, 53, 56
    variety of distributions, 46
Alpha level, 92
Anomalies
    actual density, 169, 171
    database, 174
    DATAMINE program, 183
    density and marginal product, 178
    histogram equalization, 181
    histogram normalization, 174
    implications, 180
    marginal density product, 169, 171
    marginal inconsistency, 170–172
    maxMIx and maxMIy, 179
    mean and standard deviation, 176–177
    multivariate extensions, 168
    mutual information
        contribution, 170, 172–173
    numeric values, 177–178
    optional sharpening, 182
    parameters, 182
    Parzen window method, 168
    quantities, 177–178
    scale factors, 175–176
    scale positive and negative values, 180

281

© Timothy Masters 2018
T. Masters, *Data Mining Algorithms in C++*, https://doi.org/10.1007/978-1-4842-3315-3

Get the eBook for only $5!

Why limit yourself?

With most of our titles available in both PDF and ePUB format, you can access your content wherever and however you wish—on your PC, phone, tablet, or reader.

Since you've purchased this print book, we are happy to offer you the eBook for just $5.

To learn more, go to http://www.apress.com/companion or contact support@apress.com.

Apress®

Printed in the United States
By Bookmasters